Experimental Brain Research Series 18

Guido Gainotti Carlo Caltagirone (Eds.)

Emotions and the Dual Brain

With 30 Figures and 15 Tables

Springer-Verlag Berlin Heidelberg New York
London Paris Tokyo Hong Kong

Prof. Dr. Guido Gainotti

Università Cattolica del Sacro Cuore
Facoltà di Medicina e Chirurgia "Agostino Gemelli"
Largo Agostino Gemelli, 8, 00168 Roma, Italia

Prof. Dr. Carlo Caltagirone

Clinica Neurologica, Seconda Università di Roma
Largo Umanesimo, 10, 00144 Roma, Italia

ISBN-13: 987-3-642-73398-7 e-ISBN-13: 987-3-642-73396-3
DOI: 10.1007/987-3-642-73396-3

Library of Congress Cataloging-in-Publication Data
Emotions and the dual brain / Guido Gainotti, Carlo Caltagirone (eds.).
p. cm. - (Experimental brain research series; 18)
Based on a symposium held in Rome, Nov. 27-29, 1986 under the auspices of the Istituto dell'enciclopedia italiana and dedicated to the memory of Giuseppe Moruzzi.
ISBN-13: 987-3-642-73398-7 (U.S.)
1. Emotions - Physiological aspects - Congresses. 2. Cerebral dominance - Congresses. 3. Brain - Localization of functions - Congresses. 4. Laterality - Congresses. 5. Moruzzi, Giuseppe - Congresses. I. Gainotti, Guido. II. Caltagirone, Carlo, 1948-. III. Moruzzi, Giuseppe. IV. Istituto dell'enciclopedia italiana. V. Series.
[DNLM: 1. Brain - Physiology - congresses. 2. Dominance, Cerebral - congresses. 3. Emotions - congresses. 4. Laterality - congresses. 5. Neurobiology - congresses.
W1 EX4855 v. 18 // WL 335 E54 1986] QP401.E44 1989
152.4 - dc20 DNLM/DLC for Library of Congress 89-6310

2125/3140-543210 - Printed on acid-free paper

Preface

Until the last two decades work on the neurobiology of emotions has been almost only centered on the structures forming the so-called Limbic System. This line of research was a development of the neurological models of emotion proposed in the first half of the century by Cannon, Papez and McLean and has allowed a substantial improvement of our knowledge about the neural substrates of emotions. In more recent years a quickly growing body of clinical and experimental data has, however, suggested that apart from the structures forming the Limbic System, also cerebral lateralization plays a certain role in emotional processing, since hemispheric asymmetries can be observed in comprehension, expression and (perhaps) experience of emotions. Although this notion has quickly became very popular, the exact nature of the relationships between emotions and hemispheric specialization still deserves clarification.

The goal of the symposium of which this book is a outcome and which was held in Rome, November 27-29, 1986, under the auspices of the "Istituto dell'Enciclopedia Italiana", was to try to reach a better understanding of these relationships, by gathering people who had approached this problem coming from different areas of the neuroscience, of behavioral neurology, psychology and psychopathology.

The symposium was held in memoriam of Giuseppe Moruzzi, since we felt attempt of clarification of this fascinating aspect of the neurobiology could be an appropriate way of honoring this great figure of neuroscientist.

The book is organized in three sections: the first is devoted to analyze some basic neurobiological and psychophysiological aspects of emotions; the second contains a series of reviews and of experimental papers dealing with the emotional comprehension, expression and autonomic reactivity of patients with unilateral brain damage; the third includes experimental and theoretical papers relevant to clarifying the links between psychopathology and hemispheric specialization.

The editors thank Mrs. M. Bonfante for typing several chapters and Dr. S. Conte for the invaluable work in preparing camera-ready of the book.
They also gratefully acknowledge the support of Springer-Verlag in editing this book.

Rome, Spring 1989

G. Gainotti
C. Caltagirone

Contents

List of Contributors*

---

* The address of the authors is given on the first page of each contribution
[1] Page, on which contribution commences

Remembering Giuseppe Moruzzi

G. Berlucchi
Institute of Human Physiology, Medical Faculty, University of Verona, Verona, Italy

INTRODUCTION

The Istituto dell'Enciclopedia Italiana is a most appropriate place for commemorating Giuseppe Moruzzi. During the last years of his life I had the opportunity and pleasure to see him work on many entries of the Enciclopedia del Novecento, the monumental publication of the Istituto which aims at presenting a comprehensive picture of the flow of contemporary scientific thought. I can testify that he made an enormous contribution to the success of this publication by applying, with characteristic total dedication, his wisdom and capacity for work to all stages of the production, from choice of authors to correction of galley-proofs. I also have a vivid memory of the proud satisfaction with which he secured for the Enciclopedia the collaboration of Lord Adrian (Adrian 1977) and Frederic Bremer (Bremer 1982), two great neurophysiologists who had been his influential teachers during his formative years. He felt that their writings could bring into the Enciclopedia the essence of one of his most valuable and profound spiritual roots

MORUZZI'S OUTSTANDING CONTRIBUTIONS TO NEUROSCIENCE

Moruzzi was an exceptional scientist, an exceptional scholar, and an exceptional man. These three aspects of his personality are inseparable because he succeeded in integrating them into an exemplarily coherent life, marked by a consistent harmony between his great intellectual, cultural and moral qualities. That he was an exceptional scientist is revealed by those criteria for scientific excellence whose validity he himself always upheld: the seminal discovery; the ever-lasting acquisition of facts and concepts; the experiment which is as valid today, and will be as valid tomorrow, as when it was first published; the observation which not only solves a specific scientific problem, but also teaches how to reason and make predictions in front of every natural phenomenon.

At least three of Moruzzi's papers fully meet these criteria. The first paper, published with Adrian in 1939 (Adrian and Moruzzi 1939), provided the first proof that the cortical neurons giving rise to the pyramidal tract, and by inference all corticofugal neurons, participate in the spontaneous activity of the brain by a continuous emission of action potentials, even in the absence of movement. This discovery, to which the then 28-year-old Moruzzi contributed no less than did the 1932 Nobel laureate Adrian, is still a starting point for models of the organization of the intrinsic activity of the central nervous system.

The second paper, coauthored by Moruzzi and Magoun (1949), demonstrated the activating influence of the brainstem reticular formation on the EEG, and is rightly considered Moruzzi's best work. In 1981 the journal Current Contents selected it as a "citation classic" on the basis of the more than 840 quotations received during the 1961-1981 period (Moruzzi and Magoun 1981). A perhaps less arid index of its impact on the development of the neurosciences is the almost universally accepted division of the history of these sciences into the pre- and post-Moruzzi-Magoun eras.

The third paper of major importance was published by Moruzzi in collaboration with Batini, Palestini, Rossi, and Zanchetti (Batini et al. 1959). By describing the prolonged EEG and behavioral arousing effects of pretrigeminal midpontine sections, it offered a compelling argument for the active genesis of sleep and its partial dependence on caudal brainstem structures. It also prompted a series of experimental investigations and theoretical constructions which bear out the paramount role of Moruzzi in clarifying the functional mechanisms of the sleep-wake cycle.

Moruzzi's Interest in the Problem of the Neurophysiology of Emotions

Moruzzi was a very productive scientist, and in addition to the above three famous papers he published very many studies which helped the advancement of the neurosciences in various ways and fields, particularly with respect to cerebellar physiology and pathophysiology. For space limitations, I shall consider only his contributions to the subject of this volume, the neurophysiology of emotions. As a physiologist he was chiefly interested in the basic mechanisms of animal life, and he rightly maintained that in order to begin to understand such mechanisms one ought to address issues that are simple, unitary, and possibly amenable to common biological principles. Emotional phenomena, whether introspective or behavioral, are neither simple, nor unitary, nor easily liable to biological generalization, and perhaps as a result of this Moruzzi's experimental work never tackled the problem of the neural bases of emotions directly. Yet it is a fact that his discovery with Magoun of the activating functions of the ascending reticular formation changed the course of the neurophysiological approaches to the analysis of the relations between cerebral organization and emotional life.

In 1951 - two years after Moruzzi and Magoun (1949) had obtained EEG activation by reticular stimulation, and Lindsley et al. (1949) had produced coma with lesions of the rostral pole of the reticular formation - Donald

Lindsley utilized these results for proposing a theory of emotion which has exerted a strong influence on neuroscience to this day (Lindsley 1951). The theory, called the activation theory because of the key role it attributes to EEG activation, views the hypothalamus-reticular formation complex as the central substrate which gives rise to the descending impulses controlling the somatic manifestations of emotion, as well as to the ascending corticopetal impulses mediating the related subjective experiences. According to the theory, the activity of the central substrate, which is amply open to facilitating and inhibiting influences arriving from both sensory periphery and other central nervous structures, varies in a continuous fashion between two extreme states, maximal undifferentiated emotional excitement, on the one hand, and sleep, on the other. The differentiation between discrete emotional states and the modulation of their intensity are attributed by the theory to intermediate degrees of activity of the central substrate.

The last assumption has always constituted a weak point of the theory because it is not easy to see how a mere quantitative change in the activity of the same neural system could bring about a shift, for example, from a positive to a negative emotion or viceversa. It is to Moruzzi's credit that he was one of the first to identify this weak point of the theory and to indicate a possible correction. In a scarcely known paper published in 1958 under the title "The Functional Significance of the Ascending Reticular System" (Moruzzi 1958), he suggested the possibility that the ascending reticular system is by no means homogeneous and unspecific, such that the various emotional states may result from its fractionated and differentiated activation rather than from quantitative variations of its overall diffuse energizing influence on the cortex. There is a hint in that paper at the existence of multiple and different ascending brainstem system - a concept which modern neurochemistry has shown to correspond to the reality of the various chemically differentiated ascending pathways, with substantial consequences for the understanding of the neurology of emotions.

Significant Experimental and Clinical Contributions to the Neurobiology of Emotions by the First Teacher and Two Pupils of Moruzzi

While Moruzzi did not himself perform experiments on the neurophysiology of emotions, three investigators who had strong scientific and personal ties with him made significant experimental contributions to this field. As a young medical student, Moruzzi found his first scientific mentors in two distinguished professors of the University of Parma in the 1920s and 1930s: Antonio Pensa, professor of anatomy, and Mario Camis, professor of physiology, who was responsible for introducing Moruzzi to neurophysiology. A brilliant mind, educated by Luciani, Aducco and Sherrington, Camis wrote, soon after the end of the First World War, a booklet entitled Il Meccanismo delle Emozioni (Camis 1919). The booklet did not enjoy much success and was soon forgotten, but it contained a number of first-hand experimental observations and imaginative hypotheses which are still worth our consideration. For example, it reports Camis' original experiments using the reaction-time method, a method which, as witnessed by other papers in the present volume, finds ample application in present-day analyses of emotions. In the 1950s and 1960s one of Moruzzi's pupils, Alberto Zanchetti, published with various collaborators a series of elegant investigations which underlined the chief role of the reticular formation in the triggering of sham-rage in the thalamic preparation (Zanchetti 1967). This remains one of the most interesting animal models of emotional control by the central nervous system. Another pupil of Moruzzi, the neurologist Hrayr Terzian, described with Cecotto, back in 1959, the opposite dysphoric syndromes generated by the reversible pharmacologic inactivation of one or the other hemisphere: a depressive-catastrophic reaction following left hemisphere inactivation, and a euphoric-manic reaction following right hemisphere inactivation (Terzian and Cecotto 1959). The very fact that after more than a quarter century the central subject of this volume is the functional link between hemispheric specialization and emotional processes proves the importance of that early observation, and attests to the viability of the Terzian-Cecotto assumption of a differential involvement of the two hemispheres in emotional regulation.

Moruzzi as a Scholar and Educator

Moruzzi was an authentic intellectual who added to his scientific success the great merit of being a teacher in the true sense of the word, that is one who can transmit a tradition of civilization. He played a fundamental role in the scientific and cultural revival of Italy after the tragedies of the fascist dictatorship and the Second World War, but his influence extended far beyond the boundaries of his country. In 1949, when he returned from Chicago to Pisa to take up the headship of the Institute of Physiology of the local university, he began to put into practice a program whose optimism and goodwill are made clear by a commentary he published in that year (Moruzzi 1949). In addressing a scientific American society which had invited him to describe the state of scientific research in Italy after the destructions of the war, Moruzzi gave a dispassionate picture of the moral and material poverty of his fellow citizens, but at the same time he openly revealed his hopes for the future and the task he had set for himself:

I hope that future European research work will contribute enough to justify the belief of many of us, that the old continent is still alive. Europe experienced worse times after the fall of Rome and then she alone was responsible for Western civilization. I have often wondered about the feelings of the scholars in those miserable centuries and I think that they could only be ones of despair. And yet at the beginning of the twelfth century, something happened which very few had hoped for, and men began to build cathedrals, to found universities, and to irrigate their lands. It was the dawn of the Renaissance and of our modern culture. We do not know where we are standing in the tide of civilization, but we have faith that European contributions to the humanities and sciences are not near the end.

Innumerable colleagues, collaborators, pupils, and students not only from Italy but also from many other countries, can testify that Moruzzi was always faithful to this spirit, which allowed him to create in his Institute in Pisa a spiritual atmosphere outstanding for its almost religious concept of the significance of research work, for the breadth of its cultural scope, and for its logic and methodologic rigor. Whoever was lucky enough to work and learn in that atmosphere, whoever could appreciate its direct or induced effects, whoever had a chance to meet Moruzzi and interact with him on a personal basis, cannot but be fully aware of the exceptionality of the man and his active life, and deeply feel and mourn the void left upon his death.

REFERENCES

Adrian ED (1977) Fisiologia.In: Enciclopedia del novecento, vol 2. Istituto dell'Enciclopedia Italiana, Rome, pp 1066-1068

Adrian ED, Moruzzi G (1939) Impulses in the pyramidal tract. J Physiol (London) 97: 153-199

Batini C, Moruzzi G, Palestini M, Rossi GF, Zanchetti A (1959) Effects of complete pontine transections on the sleep- wakefulness rhythm: the midpontine pretrigeminal preparation. Arch Ital Biol 97: 1-12

Bremer F (1982) Sistema reticolare ascendente. In: Enciclopedia del novecento, vol 6. Istituto dell'Enciclopedia Italiana, Rome, pp 777-800

Camis M (1919) Il meccanismo delle emozioni. Bocca, Turin, p 236 Lindsley DB (1951) Emotion. In Stevens SS (ed) Handbook of experimental psychology. Wiley, New York, pp 473-516

Lindsley DB, Bowden J, Magoun HW (1949) Effect upon the EEG of acute injury to the brain stem activating system. Electroencephalogr Clin Neurophysiol 1: 475-486

Moruzzi G (1949) Commentary. Electroencephalogr Clin Neurophysiol 1: 230

Moruzzi G (1958) The functional significance of the ascending reticular system. Arch Ital Biol 96: 17-28

Moruzzi G, Magoun HW (1949) Brain stem reticular formation and activation of the EEG. Electroencephalogr Clin Neurophysiol 1: 455-473

Moruzzi G, Magoun HW (1981) This week's citation classic. Current Contents 24 (40): 21

Terzian H, Cecotto C (1959) Su un nuovo metodo per la determinazione e lo studio della dominanza emisferica. G Psichiatria Neuropatol 87: 889-924

Zanchetti A (1967) Subcortical and cortical mechanisms in arousal and emotional behavior. In Quarton GC, Melnechuck T, Schmitt FO (eds) The neurosciences. A study program. Rockefeller University Press, New York, pp 602-614

Introduction

Features of Emotional Behavior Relevant to Neurobiology and Theories of Emotions

G. Gainotti

Institute of Neurology of the 2nd Faculty of Medicine of the University of Naples, Via Pansini 5, 80131 Napoli, Italy

INTRODUCTION

In a book designed to analyze the relationships between emotions and hemispheric specialization, a chapter outlining the most general aspects of emotional behavior and introducing some general theories about the nature of emotions must fulfill a rather precise introductory scope. A large agreement, indeed, exists about the fact that (at least in man) the term "emotion" denotes a complex functional system. Leventhal and Scherer (1987) have recently stressed this position, claiming that emotions must be distinguished from reflexes and other highly stereotyped response mechanisms, since emotion represents a phylogenetically evolved adaptation mechanism (Scherer 1984) generated by a complex, hierarchically organized, multicomponent system (Leventhal 1984). A knowledge of the internal organization of this functional system is, therefore, a necessary prerequisite to understand how emotions may be mapped into the human brain and to explore the possible links between emotions and hemispheric specialization. But if this is, in the context of the present book, the goal of a chapter outlining some general aspects of emotional behavior, then the content of the chapter must be consistent with this goal. From this point of view, an all purposes survey of current theories of emotions would not be of great usefulness, since some of the most important theories in this field are not relevant to this issue. More appropriate to our scope could, perhaps, be an attempt to analyze the functional or behavioral features which can be considered as highly characteristic of human emotion and to discuss (starting from them) the possible neurobiological correlates of these aspects of emotions and the main theories which have stressed one or another of these features of emotions. These are, therefore, the aims that we intend to tackle in the present, introductory chapter.

THE LEVENTHAL'S MODEL AS A THEORETICAL FRAMEWORK TO STUDY THE MAIN FEATURES OF HUMAN EMOTION

Although we do not necessarily subscribe to the perceptual motor theory of emotions developed by Leventhal (1974, 1979, 1984), we certainly agree with this author when he claims that:

Experimental Brain Research Series 18

(a) Every emotion is formed by many components and is based on operations arising at various levels
(b) The emotional processing system is hierarchically organized
(c) This hierarchical organization is the result of a construction made by the individual during his developmental history
Since an attempt to analyze the principal components of emotions will constitute the main scope of this chapter, I would introduce this analysis with a sketchy illustration of the structural and evolutive aspects of the emotions generating system proposed by Leventhal (1984). This author maintains that two types of hierarchies are involved in the emotional processing system. A first hierarchy concerns the relationships between the central nervous system and the peripheral machinery of the body, as the former uses information coming from the latter for constructing emotional reactions. A second hierarchy concerns several successive levels operating within the central nervous system. At the bottom of the structure is a basic set of innate neuromotor programs, each of which generates a different set of expressive reactions (and of feelings) in response to specific releasing stimuli. This basic sensory motor structure is operating at birth but, with the biological and experiential maturation of the individual, becomes partially incorporated into the more and more complex cognitive and expressive aspects of his life. Two levels of emotional processing can be distinguished, according to Leventhal, after the first periods of life:
- A schematic level, in which emotions are automatically generated in a prototypical situation (schema) and which combines the stimulus input with the subjective feelings, the expressive motor response, and the autonomic response
- A conceptual level, in which emotions are categorized, rather than felt, according to a set of abstract propositions concerning the emotion provoking situations and the rules to respond appropriately (i.e., deliberately rather than spontaneously) to emotional situations

Although the degree of involvement of these two levels in everyday emotional response may vary according to individual (developmental) and to social-cultural factors, each of these levels is involved in every emotional response.

FEATURES CHARACTERISTIC OF HUMAN EMOTION

If we pass, now, from this synthetic illustration of the Leventhal's model to the behavioral and functional features which are generally considered as characteristic of human emotions, we can say that some of these features are present at the basic sensorymotor level, whereas others are typical of the upper levels of the model. Among the former, which can be considered the "hard/wired" components of emotions, the following can be listed as highly typical of emotions:
(a) A basic set of expressive motor bodily reactions, generated by a preprogrammed central motor mechanism

(b) A basic set of specific external cues, capables of eliciting (during the first stages of neonatal developmental) the above mentioned expressive motor programs
(c) An important activation of the autonomic nervous system
(d) An increased level of arousal of the central nervous system
The last two items are concomitant with the production of the expressive motor reaction.
Among the latter features, which reflect the incorporation of the basic sensory-motor system into more evolved cognitive and executive functional systems, the following can be listed:
(e) The integration of the primitive key stimuli, capable of eliciting in a reflex-like manner the innate expressive motor programs, into more complex and variable sets of situations, which associate the external stimuli with representations and memories of past emotional experiences
(f) The partial incorporation of the emotional expressive motor programs into the whole set of communicative skills, which are deliberately used by the individual during his interpersonal interactions.
Finally, there is, separate from the features of these two levels,
(g) The subjective experience of emotions, which is certainly one of the most important components of emotional behavior, but whose nature is hotly debated and controversial. A more analytical description of each of these features follows in the next part of this section.

Basic Expressive Motor Reactions

The link between emotions and movements is already apparent in the etymology of the term "emotion" (from the latin e-motio, meaning movement from) and, indeed, emotions must be considered among the most powerful motivational determinants of behavior. They, therefore, have partly autonomous paths of access to skeletal muscles, to which they transmit patterned drive related commands. But what is perhaps even more important here is that, in addition to these command mechanisms for the skeletal apparatus, emotions also provoke highly characteristic expressive motor reactions at the level of the mimic facial muscles. A large body of data gathered in recent years by Tomkins (1962), Izard (1971, 1977), Ekman (1973, 1977), and others has shown that these patterned mimic reactions produce distinctive expressions for a limited number of basic emotions (namely happiness, surprise, fear, anger, disgust, and sadness) and suggest that these expressive reactions are innate.

These conclusions, which are consistent with Darwin's claim that facial emotional expressions are part of man's biological inheritance (Darwin 1872/1965), are based on the following data. (1) Facial expressions of emotions are recognized with high accuracy across different cultures (Ekman and Friesen, 1971; Izard, 1971). (2) They can be provoked in children from the first days of life (Field et al. 1982; Fox and Davidson 1984). (3) Blind subjects show facial expressions of emotions similar to those displayed by persons with normal vision (Eibl-Eibesfeldt 1980), an observation which demonstrates that these expressive motor reactions are innate and not learned. (4) Finally, similar distinctive facial expressions, elicited by situations provoking fear, anger, pleasure, and possibly sadness, have also been observed in nonhuman primates (Chevalier-Skolnikoff 1973; Redican 1982; Pilleri 1984). A third aspect of movement which, in addition to patterned bodily reactions and to facial expressions of emotions, is involved in emotional motor reactions concerns vocal activity. This aspect of the problem has been studied less than facial expression of emotions (Scherer 1981). It is, therefore, not clear if voice simply conveys the difference between positive and negative emotions (capturing the subject's attention and leaving the task of gathering more precise information to the inspection of facial expression) or if vocal expressions are just as emotion specific as facial expressions of emotions are (Tomkins 1962).

As for the location of structures subserving these expressive motor programs, it must be acknowledged that in animals behavioral manifestations of emotions can be obtained by stimulation of various subcortical and cortical structures (see Buck 1976; Brodal 1981; Macchi 1988), such as the hypothalamus, the amygdala, the septal nuclei, the cingulate gyrus, and the parahippocampal gyrus. However, since expressive motor reactions provoked by stimulation of the cortical areas are less consistent and more dependent on the situational context in which the stimulation takes place, it is unlikely that the command systems for the basic emotions may be located in these cortical structures. It is more likely that these systems may be located in circuits running from the hypothalamus to the amygdala and to the septal nuclei, which could be activated by the after discharge provoked by the cortical stimulation.

It is also likely that command systems for different emotions may be located in different parts of the hypothalamus and of associated structures. Panksepp (1982, and this volume) maintains, for example, that stimulation of the lateral and dorsolateral hypothalamus provokes manifestations of expectancy and anger, whereas stimulation of the anterior hypothalamus elicits fear and panic reactions. On the other hand, Delgado (1960) has obtained, by stimulating the septodiagonal area, a pleasurable response contrasting with the fear and anger reactions obtained through stimulation of the amygdala. Several reasons suggest, however, cautiousness both in attributing to the hypothalamus a pivotal role in the expression of emotions and in assuming that different parts of the hypothalamus and of the associated structures subserve the command systems for different emotions.

The first and more general reason is that even in mammals significant interspecies differences might exist in the anatomical substrata of emotions. Results of animal experiments might, therefore, be only partly relevant to understanding the anatomical organization of human emotions.

The second reason is that the nature of the behavioral responses elicited by hypothalamic stimulation is a matter of debate. Some data (Bandler 1982) suggest that emotional responses provoked by electrical stimulation of the ventral hypothalamus could be due to the activation of axons of passage and not to the excitation of local cell bodies.

The third reason which suggests cautiousness in attributing the command systems for different emotions to different parts of the hypothalamus and of the associated structures is the overlap existing between the areas whose stimulation produces a given emotional reaction and those evoking a different (usually opposite) emotional response. Sem-Jakobsen (1968) has shown, for example, with depth recording and electrical stimulation in humans, that a close spatial relationship exists between the structures provoking smiles and other positive responses, and those eliciting negative emotional reactions. Analogously it is very frequently observed in animals that the emotion elicited by stimulation of a given part of the amygdala or of the hypothalamus can change to the opposite (e.g., anger to fear) given small displacements of the electrodes, slight modifications of the parameters of stimulation, or even changes in the environmental conditions (see Macchi 1968, 1988; Brodal 1981).

The fourth consideration is that perhaps not all the command systems for different components of the drive related motor reactions are located in the same parts of the brain. McLean (1986) maintains, for example, that a difference exists between the most primitive drive-related forms of behavior (such as the fight-flight reactions) which are present in phylogenetically old species, such as the reptilians, and the family related patterns of behavior, which are characteristic only of mammals. The former would be subsumed by the hypothalamus and by the related parts of the paleostriatum, whereas the latter (namely, nursing in connection with maternal care, audiovocal communication for maternal-offspring contact, and play) would be subserved by the cingulate gyrus, which is the phylogenetically more recent part of the limbic system.

External Key Stimuli Evoking the Innate Expressive Motor Reactions

The analysis of the external stimuli automatically evoking the above mentioned basic expressive motor reactions is of theoretical interest in that the nature of these stimuli could tell us something about the nature of emotions themselves. It is note worthy from this point of view that, although in the earliest moments of life variations of the internal milieu are certainly important in the generation of the expressive motor programs, very soon these programs come under the control of specific external stimuli (Emde 1979), the most critical of which are interpersonal in nature. We do not refer here to the common observation that social stimuli, such as the voice or the face of the mother elicit a variety of orienting and interest responses in the neonate (Brazelton et al. 1974), but to the results of controlled investigations, such as the study of Field et al. (1982), which have shown that features of expressive behavior produced by adults in front of days-old infants capture the neonates' attention and elicit in them similar expressive responses. These data confirm the view that interpersonal expressive cues constitute the major group of key

external stimuli to which the neonate's expressive motor system is prepared to react automatically.
Furthermore, empirical data suggest that most of these primitive key stimuli maintain their efficacy with the incorporation of the primitive sensorimotor emotional mechanisms into more complex emotional schemata. Ohman and Dimberg (1978) have shown, for example, a continuity from the sensorimotor to the schematic processing for certain perceptual configurations (such as an angry face) and have suggested that these configurations are incorporated in a central set of emotional schemata on the basis of partly preprogrammed mechanisms.

Activation of the Autonomic Nervous System

Although activation of the autonomic nervous system is a basic component of emotions, both the qualitative aspects and the functional significance of this activation remain controversial. As for the first point, it is not clear whether each emotion has a characteristic pattern of autonomic nervous system activity, or if the autonomic responses provoked by different emotional situations differ in degree but not in quality. The hypothesis that each emotion may have a typical pattern of autonomic activity was first advanced by James (1890) and then revised by many other authors on the basis of theoretical rather than empirical reasons. More recently, Ekman et al. (1983) have also obtained experimental data in support of this hypothesis. They asked normal subjects to move combinations of facial muscles involved in the expression of fear, anger, surprise, disgust, sadness, and happiness and found that the correct production of these facial expressions induced changes of autonomic arousal which were different for each expressed emotion.

The opposite view was proposed by Cannon (1927), who claimed that the same autonomic responses occur for different feeling states. Following this line of reasoning, Schachter and Singer (1962) elaborated the "cognition-arousal" theory, which maintains that autonomic arousal is necessary but not sufficient to create the experience of a particular emotion, since it varies only in intensity but not in quality. According to these authors, autonomic arousal generates, however, an internal sensation which leads the individual to look outward for the causes of this arousal and it is this cognitive appraisal of the situation which specifies the quality (joy, fear, anger) of the feeling state.

As for the functional significance of the autonomic activation concomitant with emotional behavior, two diverging interpretations, again put forward by James (1890) and Cannon (1927), have been advanced. The James' classical "body reaction" theory maintains that the autonomic visceral response, directly provoked by the perception of the emotion eliciting situation, leads to the emotional feelings thanks to a visceral feedback mechanism.
The Cannon theory maintains, in contrast, that autonomic visceral activity is irrelevant for emotional experience, serving instead to prepare the whole organism for action. This theory could account mainly for emotions, such as fear or anger, which have common strong adrenergic effects, putting the

organism in a condition to respond quickly and strongly to the emergency situation which has provoked the emotional response.

If we pass, now, from the functional significance to the neurobiological bases of the autonomic changes concomitant with emotions, we find a situation very similar to that described while analyzing the cerebral organization of the basic emotional motor programs. Important autonomic changes can, indeed, be observed after electrical stimulation of the hypothalamus, amygdala, gyrus cinguli, parahippocampal gyrus and fronto-orbital cortex. This does not mean, however, that all these structures are equivalent with regard to autonomic arousal, since changes observed after stimulation of the hypothalamus are more strong and more consistent that those observed after stimulation of all the other cortical and subcortical structures. It is, therefore, likely that the cortical components of the limbic system may partly influence autonomic arousal through multisynaptic pathways relaying first in the amygdala and then projecting downward to the hypothalamus. The similarity between organization of the expressive motor and of the autonomic components of emotional behavior should, however, not lead us to think that autonomic and expressive motor components of emotional behavior are intrinsically related. Hess studies (reviewed by Hess 1954) had suggested that this could be true at the hypothalamic level, since hypothalamic stimulations usually provoke integrated (expressive motor, autonomic and hormonal) emotional responses. Later studies (reviewed by LeDoux 1987) have however shown that autonomic and expressive motor components of emotions can be dissociated at various levels.

Thus, the lateral hypothalamus contains neurons involved in the control of autonomic (but not of expressive motor) components of emotions (Smith and DeVito 19984) and lesions of the stria terminalis (which is one of the main pathways connecting the amygdala with the hypothalamus) blocks the motor, but not the autonomic, components of the emotional response (Hilton and Zbrozyna 1963). Furthermore, Schwaber et al. (1982) have shown that the central nucleus of the amygdala has direct connections to brainstem autonomic regions, such as the nucleus of the solitary tract, and could, therefore, bypass the hypothalamus in controlling autonomic aspects of emotions. Finally, recent findings obtained with the classical conditioning procedure and also reviewed by LeDoux (1987) suggest that the hypothalamus might be more critically involved in the autonomic than in the expressive motor aspects of emotions.

Increased Level of Central Nervous System Arousal

One of the main characteristics of emotional experience consists in the high level of alertness and of neurophysiological arousal which typically accompanies emotions. This increased level of central nervous system arousal can have very different consequences at the behavioral level. When the emotion is moderate, it can amplify the interest and the attentional capacities of the subject, allowing him to improve his analysis of the relevant incoming sensory informations. When, however, the emotional components of a

situation are overriding, then the excessive and generalized excitement of the central nervous system has disorganizing effects, hampering the subject's ability to tackle the emotion eliciting situation.
From the neurophysiological point of view, the increased level of alertness is accompanied by a change of the EEG resting pattern, with disappearance of the synchronized alpha activity and appearance of a slow amplitude and fast activity beta rhythm. These behavioral and neurophysiological observations led Lindsley (1951) to formulate his well known activation theory. According to this theory, electrical stimulation of the upper part of the reticular formation not only evokes an EEG desynchronization similar to that occurring during spontaneous arousal, but also provokes a descending activation of autonomic and expressive motor structures.
The concomitance of ascending and descending activation, with behavioral and EEG signs of cortical arousal and with expressive motor and autonomic manifestations of emotional behavior, is particularly strong, according to Lindsley, when the electrical stimulation is applied at the hypothalamic level. A discharge of the hypothalamic structures could, therefore, account both for the central and for the peripheral aspects of emotional behavior.

Memory and Cognitive Components of Emotions

Although the elementary sensorimotor and vegetative components of the emotion generating mechanisms are likely to be genetically inscribed into the central nervous system, the incorporation in this mechanism of specific situations (which constitute the source of particular emotional experiences) is the result of a learning process. According to Gray (1972), who views emotions as conceptual states of the nervous system that are used to categorize ogoing events, emotions are produced by the association between a situation and a reinforcing event (or a stimulus which in the subject's previous experience has been followed by a reinforcing event). Similar to Gray's conceptualization is Leventhal's construct of the schematic emotional processing. Schemata are considered by Leventhal (1984) as concrete memory representations of specific perceptual, motor, and subjective feelings experienced during emotional encounters with the environment. Schemata organize the emotional experience since they form prototypical instances of whole classes of emotional events; they link objects to feelings and create emotional blends which render the emotional life of the subject richer and more complex. According to Leventhal, emotional schemata, created through a mechanism of conditioned reaction, are automatically activated by external events similar to those stored in the emotional representation and do not require the participation of an abstract, propositional processing. They must, therefore, be distinguished from the highest levels of the emotional processing, which are conceptual and volitional rather than automatic in nature and which are based on the highest cognitive and ethical aspects of the personality rather than on a mechanism of conditioned response.

The distinction between schematic and conceptual processing of emotional information is not explicitly accepted by cognitive theories of emotions, which put the greatest emphasis on the process of cognitive appraisal of the emotion provoking situation, without necessarily specifying if this appraisal is completely or only in part controlled. Different cognitive theories, such as those put forward by Lazarus (1982, 1984) and Frijda (1984), stress, in turn, different aspects of the process of appraisal. Frijda, who acknowledges that most appraisal processes are, indeed, automatic, suggests, for example, that the different emotions might be considered as different outcomes of primary and secondary appraisals, where the term "primary" refers to the assessment of whether event outcomes will be positive or negative and the term "secondary" refers to the appraisal of the coping strategies which are feasible and likely to succeed. Instead of drawing on these differences between the various cognitive theories, we prefer to turn to the possible neurobiological foundations of this aspect of human emotions. From the neurobiological point of view, the capacity of attributing emotional value to different sets of stimuli and situations presupposes the development of two important anatomical prerequisites:

(a) A set of corticolimbic connections, integrating the external perceptual information with the intrinsic (expressive motor, autonomic, and subjective) components of the emotions generating mechanism

(b) A memory system allowing the acquisition and storage of these integrated emotional experiences.

In recent years, important advancements have been made in the clarification of both these issues, since direct connections have been demonstrated between unimodal and multimodal sensory association areas and some components of the limbic system, and since the hippocampal formation has been identified as the limbic structure mainly involved in memory processes. The limbic structures which are more strongly and more directly interconnected with the sensory association areas are the cingulate cortex (Pandya and Seltzer 1982; Pandya and Yeterian 1984), the parahippocampal gyrus (Witter and Groenewegen 1986), and the amygdala (Doty, this volume). In particular, the amygdala could perhaps be viewed as the major link between the cortical areas devoted to this analysis of the external sensory information and the deep brain structures critically involved in the "hard-wired" aspects of emotional behavior (Doty, this volume). It could, therefore, play a critical role in attaching emotional and motivational significance to sensory stimuli. On the other hand, the parahippocampal gyrus, which is strongly interconnected with the hippocampus (Van Hoesen 1982), the amygdala, and the sensory association areas, could play a pivotal role in linking together perceptual, memory, and emotional aspects of a given situation.

Volitional Control and Deliberate Use of the Expressive Motor Programs During Social Interactions

The same adaptive mechanism which operates at the "cognitive" level, incorporating the primitive key stimuli into complex perceptual and symbolic configurations, also operates at the "executive" level, inserting (or trying to insert) the emotional expressive motor programs into the whole set of expressive and communicative instruments which the individual can use during social interactions. Two aspects of the problem can be distinguished. The first concerns the capacity to inhibit a spontaneous emotional expression, elicited by an appropriate stimulus; the second concerns the capacity to convincingly simulate an emotion which is not felt by the subject.

Both these aspects of intentional control are usually learned in a cultural context characterized by systems of rules about "who can show what emotion to whom and when they can show it" (Ekman 1984). Examples of display rules in Western culture are, for example, that people should look pleased and smile, but not express anger, during social activities such as dinners or cocktail parties, that men should not cry and women should not express anger during public events and so on. Since these display rules vary with social classes and across cultures, they could be responsible for the frequent observation of cultural differences in facial expression of emotions. Ekman (1984) has elegantly demonstrated that this is true, showing in a single experiment that persons belonging to different cultures and exposed to the same emotional stimuli show identical facial expressions in private, but very different expressions in public. In the first situation, spontaneous emotional expressions result from elicitation of biologically based, universal expressive motor programs, whereas in the second situation the controlled expressive behavior results from the intentional management of the expressive motor programs according to culturally learned (often overlearned) display rules. The ability to manage an expression according to display rules varies according to the force of the learned habits and to the intensity of the emotion eliciting situation.

Some aspects of the capacity to inhibit involuntary expressions and to convincingly simulate an emotion that is not felt have been studied by Ekman and Friesen (1974 and 1982). These authors have shown, for example, that some persons are able to conceal completely their reactions to stress inducing films, so as to avoid detection of possible facial leakages even by microscopic measurements (Ekman and Friesen 1974). They have also shown that, although emotional expressions can be convincingly simulated, a number of markers (such as the duration or the coordination of muscular movements) can allow detection of whether a facially expressed emotion is felt or not (Ekman and Friesen 1982). Ekman (1984) has also suggested that a different degree of intentional control might be attained in different channels of emotional expression and that it may be harder, for example, to inhibit signs of emotion in the voice than in facial expression or in body movements. The existence of subtle but consistent differences between felt and simulated emotional expressions and the different difficulty met in inhibiting signs of emotions in different expressive channels suggest that partly different mechanisms subserve spontaneous expressive motor reactions and volitional, deliberate use of the expressive motor systems. The neurological substrate of the deliberate use of the mimic facial muscles according to learned display rules is represented by the prefrontal cortex and by the pyramidal system. The former elaborates the plan of movement considered as most appropriate to the interpersonal situation, whereas the latter renders it executes. During the execution phase, commands emanating from the primary motor strip move to the brain stem through the pyramidal tract, reaching in particular the lower motor neurons of the facial nerve. As for the brain mechanisms involved in spontaneous expressive reactions, different pathways could transmit to the nucleus of the facial nerve and to the spinal cord the orders of movement emanating from the limbic command systems. One of these paths could pass through the central gray region of the midbrain, since lesions of this structure block the expression of emotional responses elicited by hypothalamic stimulation (Glusman 1974). Another path could pass through the ventral striatum and the ventral pallidum, since these structures are strongly interconnected with the limbic system (Nauta 1986) and seem to play an essential role in the patterned organization of drive related motor responses (McLean 1981, 1986).

The Subjective Experience of Emotions

Although the features of emotional behavior that we have described in previous sections are highly characteristic of human emotions, none of them necessarily concerns the problem of feelings (or subjective experience) which, for many people, are the essence of emotions. The fact is that the nature of feelings can only be inferred in animals (see Panksepp, this volume) and can hardly be studied in humans; accordingly, different theories give different interpretations of the mechanisms underlying emotional experience. We have tried to summarize in Table 1 some of these theories linking emotional feelings to some of the features of emotional behavior that we have just discussed.

A first class of theories maintains that feelings result from a feedback coming from output components of the emotional behavior. These theories differ, however, in the identification of the kind of expressive response which precisely provokes a particular type of emotional experience. The classical James (1890) theory assumed that the visceral change, rather than the somatic motor expression, was the component of the body reaction (directly following the perception of the exciting fact) which provided the most important feedback for emotional experience. The facial feedback hypothesis (Tomkins 1962; Izard 1971) disagrees, however, and states that visceral changes cannot be the main source of feelings, since viscera are relatively insensitive and produce similar responses for different emotional states (Cannon 1927). A better candidate for the feedback hypothesis could be the facial expressive apparatus since (a) distinctive and probably innate facial expressions have been described for the basic emotions, and (b) according to Gellhorn (1964) sensory information coming from the facial muscles can easily reach the hypothalamic nuclei. An exhaustive discussion of the facial feedback hypothesis is provided by Lanzetta and McHugo (this volume).
An intermediate position between the visceral changes and the facial feedback hypothesis has been proposed by Allport (1924), who has maintained that emotional experience may result from the combined effect of autonomic and somatic responses. According to this author, the global affective influence of the autonomic arousal would be structured by the more specific information coming from the motor expressive apparatus.

Table 1. Theories linking emotional feelings to the main features of emotional behavior

Features of Emotional behavior taken into account	Theory of emotions
PERIPHERAL THEORIES	
Feedback coming from the visceral change	Body reaction theory (James 1890)
Feedback coming from the response of the somatic expressive apparatus	Facial feedback hypothesis (Tomkins 1962) (Izard 1971)
Combined effect of autonomic and somatic responses	Allport (1924)
CENTRAL THEORIES	
Cognition-aurousal theory	
Combined effect of autonomic arousal and cognitive factors	Schachter and Singer (1962)
Cognitive theories	
Cognitive appraisal of the emotional event (personal meaning and possibility of coping with it)	Stimulus evaluation checks theory (Scherer 1984) (Lazarus 1982 and 1984) (Frijda 1984)
"Central" neurobiological theories	
Activation of the cortical areas by signals coming from the subcortical emotions command systems	Cannon (1927) Papez (1937)
Cortical appreciation of changes undergone in the subcortical structures of the limbic system	(Panksepp this volume)
Activation of the highest cortical areas by impulses emanating from the midbrain reticular formation	Activation theory (Lindsey 1951)
"Central" structural theory	
Competition within the central motor program for emotional expression between schematic and conceptual emotion systems	Leventhal (1984)

If the peripheral theories of emotions suggest that emotions are generated in the body and that only the reason is built within the brain (see Panksepp, this volume), central theories maintain, in contrast, that emotions too are generated in the brain (or by cognitive processes elaborated in the brain).
A bridge between central and peripheral theories is provided by the cognition-arousal theory (Schachter and Singer 1962), which maintains that the sources of emotional experience come in part from the body and in part from the brain. To be sure, Schachter and Singer (1962) recognize, as Allport (1924) did, the role played by autonomic arousal in providing a strong affective tone to the emotional experience. In contrast to Allport, however, they maintain that the affective tone provided by autonomic arousal is not structured by other peripheral, although more specific, information, but by a central cognitive factor, consisting in the interpretation of the external event which has caused the internal arousal. The emotional experience is considered in this theory as the product of an interaction between physiologic autonomic arousal (determining the intensity but not the quality of the feelings) and the cognitive interpretation of the arousing event giving different meanings and feelings to the internal state of activation. Since the cognition-arousal theory has not received strong support from experimental investigations which have tried to put it to the test (see Cotton 1981; Leventhal 1984 for reviews) subsequent cognitively oriented authors have stressed only the cognitive rather than the autonomic components of the emotional experience. From the point of view of the present book, the most relevant of these theories is perhaps the one proposed by Scherer (1984), since this author has tried to specify in detail which cognitive processes give rise to particular emotions and how these cognitive processes might work. According to Scherer, specific emotions are brought into play by a sequence of five "stimulus evaluation checks" (namely "novelty" check, "intrinsic pleasantness" check, "goal/need significance" check, "coping potential" check and "norm/self compatibility" check) performed by mechanisms that continually scan the ongoing environment.

A description of the first two checks will give an idea of how this mechanism could work and of what emotions could result from the outcome of each check. The "novelty" check determines whether there is a change in the pattern of internal or external stimulation. The emotion of "surprise" could be seen as the positive outcome of this check. The second check concerns the "intrinsic pleasantness" of the situation and determines whether a stimulus event is pleasant or unpleasant. In the first case the resulting emotion is joy (with induction of approaching tendencies); in the second case it is distress (with induction of avoiding tendencies). Other common emotions are viewed by Scherer as less simple and resulting from specific outcomes of series of checks.

Other cognitive theories have also considered the diverse emotions as resulting from the outcome of several appraisals (see, for example Frijda 1984), but I do not think it necessary to discuss them here in any detail since they are not directly relevant to the present chapter.

More important from a neurobiological point of view are theories using the term "central" in a neurological rather than in a cognitive sense, to suggest that the source of the emotional experience is located in the cortical components of the central nervous system. Three different central neurological theories have been advanced. The first one assumes that signals emanating from the emotion command systems determine subjective emotional feelings when they reach the cortical areas. The second theory maintains that the highest cortical areas cognitively appreciate emotions by reading the changes taking place in the subcortically located emotion control systems. The third theory suggests that activation of the highest cortical centers by impulses emanating from the midbrain reticular formation provokes emotional feelings.

The first theory was put forward by Cannon (1927), who thought that emotions were generated in the thalamic nuclei and that signals emanating from the thalamus elicited both somatic expressions of emotions (when they reached the peripheral expressive apparatus) and subjective feelings (when they reached the cerebral cortex). This theory was afterwards translated into anatomical terms by Papez (1937). This author assumed that emotional stimuli, after having elicited an emotional response at the hypothalamic level, could reach the cingulate gyrus through the mammillo-thalamic tract and the anterior nucleus of the thalamus. The extensive connections linking the cingulate gyrus with the associative cortical areas could, then provoke the arousal of subjective emotional feelings.

The second theory has been proposed by Panksepp (this volume), who views the cortical appreciation of the hypothalamic changes as a revised and more convincing version of the James-Lange classical hypothesis.

Finally, the third theory is part of the classical activation theory (Lindsley 1951), viewing central and peripheral aspects of emotions as resulting from ascending and descending effects of a state of activation of the midbrain reticular formation.
A still different type of "central" theory has been recently put forward by Leventhal (1984) within the context of the perceptual motor theory. Leventhal assumes that emotion might arise from the outcome of a competition existing within the central motor program between schematic and conceptual emotion processing systems. When the automatic type of response prevails, the expressive motor reaction is accompanied by a corresponding emotional feeling. When, in contrast, the expressive motor response is due to the action of the propositional, volitional system, then the sensation of movement is not accompanied by a corresponding emotional experience.

THE ANALYSIS OF THE FEATURES OF EMOTIONAL BEHAVIOR AND THE SEARCH FOR FACTORS UNDERLYNG LATERALITY EFFECTS FOR EMOTIONS

From the point of view of the specific subject of this book, however, the major interest of an analysis of the most relevant features of emotional behavior does not lie in the neurobiological and theoretical problems that we have synthetically considered till now, but rather resides in the opportunity that an analytical approach offers for a more systematic pursuit of the factors which could underlie laterality effects for emotions. In recent years, the growing evidence suggesting a link between emotions and hemispheric specialization has, indeed, induced several authors to search for functional or structural asymmetries basically underpinning laterality effects for emotions. This search, however, has been conducted in a rather unsystematic manner, owing to the difficulty of precisely defining the nature of interhemispheric differences in the treatment of emotions and to the lack of a frame of reference which could assist in this activity. The list of the main features of emotional behavior could be used to fill the second gap and help to understand if the laterality effects for emotions concern all the aspects of emotional behavior or only some of them. In many chapters of the present book this question will be approached from a data-oriented perspective, i.e., reporting results of investigations conducted on patients with unilateral brain lesions in which one or few of the above mentioned aspects of emotional behavior had been thoroughly investigated. Thus, Pizzamiglio and Mammucari focus their attention on possible hemispheric asymmetries in spontaneous and deliberate facial expressions of emotions; Etcoff reports results of research relevant to the issue of hemispheric differences in comprehension and cognitive appreciation of emotional material; Caltagirone and coworkers go deeply into the study of asymmetries in autonomic arousal and of the links existing between autonomic and expressive motor responses; and Gainotti provides evidence suggesting that in patients with right hemisphere damage the basic defect could lie at the level of the subjective experience of emotions. Other chapters of the book approach a similar question from a more theoretically oriented perspective, stressing preferentially one or another component of human emotion in models accounting for hemispheric asymmetries in emotional processing. Thus, Tucker puts the major stress on asymmetries in cortical activation, suggesting a different lateralization of systems subserving respectively the functions of tonic, action-oriented "activation" and of phasic, perception-oriented "arousal". On the other hand, Panksepp observes that In animal studies laterality effects for emotions have been obtained only in tasks requiring an important effect of learning. Finally, Tucker, in an alternative (or complementary) intorprotation to the activation-arousal hypothesis, suggests that the sort of cognition typically associated with the two halves of the brain could explain the different involvement of the right and left hemisphere in emotional situations (but Feyereisen questions the thesis, viewing the right and left hemispheres as based respectively on a sequential-analytical vs a syncretic-holistic mode of stimulus processing). Thus, the frame of reference of the different features involved in emotional behavior can help both to gather new empirical data and to construct theoretical models aimed at explaining laterality effects for emotions. Another important factor to be systematically taken into account, both in experimental research and in the construction of theoretical models, in the articulation within each hemisphere among the structures subserving various aspects of emotional behavior. A first step in this direction is made in this volume by Kinsbourne and Rizzolatti, who rightly stress the difference between the role played in emotional behavior by the anterior and the posterior parts of the brain, but data surveyed in the present chapter clearly show that other dissociations could be suggested between medial and lateral parts of the cerebral cortex or between cortical and subcortical structures. A further distinction within the corticolimbic connections has been suggested by Bear (1983), who has distinguished a "ventral" (or temporofrontal) and a "dorsal" (or parietofrontal) visuolimbic system. Bear assumes that the former might be a "foveal" system, involved in object recognition and relevant for acquisition and storage of visuoemotional associations, whereas the latter might be a "spatial surveillance" system, playing a critical role in the detection of emotionally relevant stimuli and orienting toward them the foveal emotional learning system.

An important theoretical problem, which is raised by Kinsbourne's and Rizzolatti's contributions, resides in evaluating if intrahemispheric effects are similar within the right and left hemispheres, or if important interactions exist between lateralization and intrahemispheric organization of functions involved in emotional behavior. If the first hypothesis is correct, emotions should have a quantitatively different but qualitatively similar organization within the right and left hemisphere, with a prevalence of functions of control over emotional expression in the anterior frontal areas and of functions of emotional comprehension and appraisal in the posterior cortical areas. If the second hypothesis is correct, this general plan should no longer hold for emotional behavior and we could have, for example, a situation similar to the "diagonal pattern" organization proposed in this volume by Kinsbourne or to Davidson's (1984) model, proposing important interactions and dynamic interplays between right and left, anterior and posterior brain areas. A problem with these models consists, in my opinion, in the fact that they are often very powerful and difficult to falsify, but even in this case the distinction between different components of emotional behavior could be of great usefulness for two main reasons. The first is that data gathered following this frame of reference could provide a sound inventory of database against which to match the models themselves. The second is that the distinction between the main features of emotional behavior could increase the number of predictions that could be made starting from the models and could, therefore, allow the checking of these theories by means of complex and articulated experimental designs.
We would conclude these introductory remarks by noting that a goal similar to the one proposed in the present introduction had been devised some years ago by Scherer (1984) in a similar, although more theoretically oriented, article. Scherer concluded his paper by noting that "in the long run, fragmentary studies of the individual components of emotions, such as physiological arousal, facial expression or subjective feeling states in isolation, are unlikely to yield a major contribution to our understanding of the process of emotion. A long-term series of studies in which the phenomenon of emotion is studied in terms of all the components and levels described earlier using modern methodology is necessary". Not only do we subscribe to Scherer's statements, but we also think that some research programs reported in this volume (such as Etcoff's investigations or the integrated series of studies

that I have performed in collaboration with Caltagirone, Mammucari, Pizzamiglio, and Zoccolotti) provide the first preliminary attempts to orient, in the direction indicated by Scherer, the study of the relationships between emotions and hemispheric specialization.

References

Allport TH (1924) Social psychology. Houghton Mifflin, Boston

Bandler RJ (1982) Induction of "rage" following microinjections of glutamate into midbrain but not hypothalamus of cats. Neurosci Lett 30: 183-188

Bear DM (1983) Hemispheric specialization and the neurology of emotion. Arch Neurol 40: 195-402

Brazelton TB, Koslowski B, Main M (1974) The origins of reciprocity: The early mother-infant interaction. In: Lewis M, Rosenblum LA (eds) The effect of the infant on its caregiver. Wiley, New York

Brodal A (1981) Neurological anatomy in relation to clinical medicine, 3rd ed. Oxford University Press, New York

Buck R (1976) Human motivation and emotion. Wiley, New York

Cannon WB (1927) The James-Lange theory of emotion: a critical examination and an alternative theory. Am J Psychol 39: 106-124

Chevalier-Skolnikoff S (1973) Facial expressions of emotion in nonhuman primates. In: Ekman P (ed) Darwin and facial expression: a century of research in review. Academic, New York

Cotton JL (1981) A review of research on Schachter's theory of emotion and the misattribution of arousal. Europ J Soc Psychol 11:: 365-397

Darwin C (1872, 1965) The expression of the emotions in man and animals. John Murray, London 1872. University of Chicago Press, Chicago 1965

Davidson RJ (1984) Affect, cognition, and hemispheric specialization. In: Izard CE, Kagan J, Zajonc RB (eds) Emotion, cognition, and behavior. Cambridge University Press, Cambridge, pp 320-365

Delgado JMR (1960) Emotional behavior in animals and human. Psychiatric Research Reports 12: 259-266

Ekman P (1973) Cross-cultural studies of facial expression. In: Ekman P (ed) Darwin and facial expression: a century of research in review. Academic, New York

Ekman P (1977) Biological and cultural contributions to body and facial movement. In: Blacking J (ed) Anthropology of the body. Academic, London

Ekman P (1984) Expression and the Nature of Emotion. In: Scherer K, Ekman P (eds) Approaches to emotion. Erlbaum, Hillsdale

Ekman P, Friesen WV (1971) Constants across culture in the face and emotion. J Personality Soc Psychol 17: 124-129

Ekman P, Friesen WV (1974) Detecting deception from body or face. J Personality Soc Psychol 29(3): 288-298

Ekman P, Friesen WV (1982) Felt, false, and miserable smiles. J Nonverbal Behav 6(4): 238-252

Ekman P, Levenson RW, Friesen WV (1983) Autonomic nervous system activity distinguishes between emotions. Science 221: 1208-1210
Emde RN (1979) Levels of meaning for infant emotion: A biosocial view. Paper presented at a seminar on emotion and cognition. Social Science Research Council, San Francisco
Field TM, Woodson R,, Greenberg R, Cohen D (1982) Discrimination and imitation of facial expression by neonates. Science 218: 179-181
Fox NA, Davidson RJ (1984) Hemispher substrates of affect: A developmental model. In: Fox NA, Davidson RJ (eds) The psychology of affective development. Erlbaum, Hillasdale
Frijda NH (1984) Toward a model of emotion. In: Spielberger CD, Sarason IG, Defares P (eds) Stress and anxiety, vol 9. Hemisphere, Washington, pp 3-16
Gallhorn E (1964) Motion and emotion: the role of proprioception in the physiology and pathology of the emotions. Psychol Rev 71: 457-472
Glusman M (1974) The hypothalamic "savage" syndrome. Res Publ Assoc Res Nerv Ment Dis 52: 52-90
Gray JA (1972) The structure of the emotions and the limbic system. In: Physiology of emotion and psychosomatic illness. CIBA Symposium No 8. Elsevier, Amsterdam
Hess WR (1954) Diencephalon: autonomic and extrapiramidal functions. Grune and Stratton, New York
Hilton SM, Zbrozyna AW (1963) Amygdaloid region for defence reactions and its efferent pathway to the brain stem. J Physiol Lond 165: 160-173
Izard CE (1971) The face of emotion. Appleton-Century-Crofts, New York
James W (1890) The principles of psychology, 2 vol. Holt, New York
Lazarus RS (1982) Thoughts on the relations between emotion and cognition. Am Psychol 37(9): 1019-1024
Lazarus RS (1984) On the primacy of cognition. Am Psychol 39: 124-129
LeDoux JE (1987) Emotion. In: Mountcastle VB, Plum R, Geiger SR (eds) Handbook of Physiology, sect I: The nervous system, vol V, Higher functions of the brain, Part I. American Physiological Society, Bethesda, Maryland, pp 419-459
Leventhal H (1974) Emotions: a basic problem for social psychology: classic and contemporary integrations. McNally, Chicago
Leventhal H (1979) A perceptual-motor processing model of emotion. In: Pliner P, Blankenstein K, Spigel IM (eds) Perception of emotion in self and others, vol 5. Plenum, New York
Leventhal H (1984) A perceptual motor theory of emotion. In: Berkowitz L (ed) Advances in experimental social psychology, vol 17. Academic, New York, pp 117-182
Leventhal H, Scherer K (1987) The relationship of emotion to cognition: a functional approach to a semantic controversy. Cognition Emotion 1: 3-28
Lindsley DB (1951) Emotions. In: Stevens SS (ed) Handbook of experimental psychology. Wiley, New York, pp 473-516
Macchi G (1968) Aspetti della organizzazione anatomo-funzionale delle strutture rinencefaliche. Acta Neurol 23: 964-1021
Macchi G (1987) Substrati anatomici del comportamento emozionale.In: Caltagirone C, Gainotti G (eds) Emozioni e specializzazione emisferica. Istituto dell'Enciclopedia Italiana, Rome

McLean PD (1961) Le systeme limbique du point de vue de la self-protection et de la conservation de l'esp}ce. In: Alajouanine T (ed) Physiologie et pathologie du rhinencephale. Masson, Paris, pp 111-126
McLean PD (1986) Culminating developments in the evolution of the limbic system: the thalamo-cingulate division. In: Doane BK, Livingstone KE (eds) The limbic system: functional organization and clinical disorders. Raven, New York
Nauta WJH (1986) Circuitous connections linking cerebral cortex, limbic system and corpus striatum. In: Doane BK, Livingstone KE (eds) The limbic system: functional organization and clinical disorders. Raven, New York
Ohman A, Dimberg V (1978) Facial expressions as conditioned stimuli for electrodermal responses: a case for preparedness? J Personality Soc Psychol 36: 1251-1258
Pandya DN, Seltzer B (1982) Association areas of the cerebral cortex. Rends Neurosci 5: 386-390
Pandya DN, Yeterian EH (1984) Proposed neural circuitry for spatial memory in the primate brain. Neuropsychologia 22: 109-122
Panksepp J (1982) Toward a general psychobiological theory of emotions. Beh Brain Sci 5: 407-467
Papez JW (1937) A proposed mechanism of emotion. Arch Neurol Psych (Chicago) 38: 725-743
Pilleri G (1984) Movements related to the preservation of the species and social instincts appearing as neurological symptoms in degenerative dementias. In: Pilleri G, Tagliavini F (eds) Brain pathology. Brain Anatomy Institute, Berne
Redican WK (1982) An evolutionary perspective on human facial displays. In: Ekman P (ed) Emotion in the human face, 2nd ed. Cambridge University Press, New York
Schachter S, Singer J (1962) Cognitive, social and physiological determinants of emotional state. Psychol Rev 69: 379-399
Scherer KR (1981) Speech and emotional states. In: Darby J (ed) Speech evaluation in psychiatry. Grune and Stratton, New York
Scherer KR (1984) On the nature and function of emotion: A component process approach. In: Scherer KR, Ekman P (eds) Approaches to emotion. Erlbaum, Hillsdale, pp 293-317
Schwaber JS, Kapp BS, Higgins GA, Rapp PR (1982) Amygdaloid and basal forebrain direct connections with the nucleus of the solitary tract and the dorsal motor nucleus. J Neurosci 2: 1424-1438
Sem-Jakobsen CW (1968) Depth-electroencephalographic stimulation of the human brain and behavior. Thomas, Springfield
Smith OA, DeVito JL (1984) Central neural integration for the control of autonomic responses associated with emotion. Annu Rev Neurosci 7:: 43-65
Tomkins SS (1962) Affect, imagery, consciousness,, vol I. The positive affects. Springer, New York Berlin Heidelberg
Van Hoesen CW (1982) The parahippocampal gyrus. New observations regarding its cortical connections in the monkeys. Trends Neurosci 5: 345-350
Witter MP, Groenewegen HJ (1986) Connections of the parahippocampal cortex in the cat. III. Cortical and thalamic efferents. J Comp Neurol 252: 1-31

Neurobiological and Psychophysiological Aspects of Emotions

The Psychobiology of Emotions: The Animal Side of Human Feelings

Jaak Panksepp
Department of Psychology, Bowling Green State University, Bowling Green, Ohio 43403, USA

INTRODUCTION

During the past decade the analysis of emotions has become a major research area in human experimental psychology and cognitive neuropsychology. A comparable fervor has not evolved in the many specialties which study animal behavior from a biological perspective. Except for a few pockets of research activity, the study of emotions is generally shunned by both animal behaviorists and neuroscientists (partially because funding, at least in the United States, is scarce for such work, and partially because of the many conceptual problems which still burden the area). One need only scan the largest repository of new findings in brain research - the abstracts of the Annual Meeting for the Society of Neuroscience (U.S.) - to appreciate how little work is being conducted in the psychobiology of emotions. Hence, there is relatively little progress to summarize since I last detailed the relevant brain literature on emotions (for past reviews see Panksepp 1981, 1982, 1985, 1986a,b), except for some substantial advances in our understanding of how "fear" is elaborated via the central nucleus of the amygdala (e.g., see Iwata et al. 1986 also see a forthcoming review on fear mechanisms in the brain by Panksepp et al. 1989). Accordingly, I plan no new literature review for this chapter, but rather my goal is to i) discuss the general conceptual basis for more work on the organization of emotionality in the animal brain, ii) discuss how psychobiological data from animals relates to human emotionality, and iii) relate the animal data, mainly on subcortical emotional processes, to recent laterality research in humans.

BRAIN SOURCES OF EMOTIONALITY

Localized electrical stimulation of the neocortex and anatomically related zones of the basal ganglia (i.e., the dorsal or somatic striatum) and thalamus yield few coherent behavioral changes in experimental animals. If the stimulation is applied to more ventral areas of the brain, including limbic cortex, hypothalamus, ventral (i.e., visceral) striatum, and related mesencephalic areas, coherent behavior patterns reflective of apparent emotional states, are commonly observed. Although a large variety of specific

Experimental Brain Research Series 18

behavioral changes can be electrically elicited, they do appear to fall into a limited number of natural categories: Animals may exhibit various forms of aggressive behaviors suggesting the animal is angry; at nearby sites animals cower and/or flee as if afraid; at others, animals plaintively vocalize as if lost or abandoned; at others, they vigorously explore and investigate their surroundings as if curiously searching for something (if given appropriate goal objects, such animals commonly begin to eat, drink, gnaw, copulate, or exhibit some other type of species-typical appetitive behavior such as predation in response to the brain stimulation (see Panksepp, 1981 for summary). When these parts of the brain are damaged, animals exhibit reliable deficits (albeit not the total elimination) of the above emotive behaviors - reduced rage, fear, separation distress, and appetitive eagerness. It thus seems that the basic impulse for emotionality emerges from circuits which course through the diencephalon, especially the hypothalamus (although separation distress and play circuitry appear to be heavily represented in the thalamus). These emotive behaviors appear to be accompanied by internal affective states, since animals exhibit characteristic approach and escape behaviors toward these types of brain stimulation, and humans typically have the expected affective experiences (for review see Panksepp, 1985). Although we may never be able to measure the subjective experiences of other animals, the study of these circuits provides the most direct avenue of inquiry into basic emotional processes elaborated in the mammalian brain, including that of humans.

There is neither an agreed upon definition of emotions nor a generally accepted methodology by which to study such processes. Since the concept of emotion is ultimately derived from subjective experiences of humans, and since we cannot symbolically share such states with other animals (indeed, clear emotional communication between humans is difficult enough), emotional concepts are rarely used in behavioral brain research. Indeed, behavioral scientists during the middle third of this century chose to conceptually cripple themselves (by eliminating talk about internal causes) in the hope that the resulting empirical rigor would help establish a lasting and credible science of behavior. Accordingly, a tradition evolved in experimental psychology of avoiding intervening psychological constructs which could not be visually observed. In doing this, the scientific study of behavior cut itself off from many natural brain phenomena which remain in need of explanation. The failure to bring the concept of emotion under close empirical scrutiny in the psychobiology laboratory remains a serious obstacle to understanding how the brain governs behavior. Although the tradition of putting the behavioral cart before the neuropsychological horse facilitated methodological rigor, it impoverished our thinking about the nature of intervening processes in the brain. It seems unlikely that functional brain organization can be adequately revealed without using emotional constructs. Conversely,conceptually-liberated neurotheoretical approaches to the study of the animal brain may yield subtle new ways of looking at animal behavior and concurrently provide a substantive decoding of the underlying nature of human emotionality. Without critical psychobiological studies in lower animals, it seems unlikely that we can unravel the deep neurobiological nature of human emotions. Still, because of past traditions, it remains all too common for cognitive scientists, clinical psychologists and philosophers to look askance at such a proposition.

This strategy, which might be called "comparative neurophenomenology" is largely premised on the fact that the basic organizational plan for emotionality resides in subcortical areas of the brain which are homologously organized in all mammals, and there is little behavioral evidence that the crowning glory of brain evolution, the neocortex, has much to do with the construction of emotional feelings. For instance, adult rats which have been decorticated within the first week of birth are noteworthy for their outward normality (Whishaw and Kolb 1983) - these animals exhibit as much species-typical (and class-characteristic) "emotive energy" as neurologically intact animals. I have used such animals in my laboratory classes, and untutored students are routinely unable to guess which animals are decorticates and which are intact. Indeed, the decorticates, because of their apparently unfettered emotive energies, are deemed to be neurologically intact more often than not. From observing such animals, one is tempted to conclude that the role of the cortex in emotionality is little more than a neural loom upon which the actual details of emotional and cognitive experiences are woven but which, in and of itself, does not instigate the feeling of emotions (albeit it surely can initiate the chain of neural events that lead to emotions). Limbic cortex can probably elaborate intrinsic perceptual functions which trigger emotions, but recent evidence also indicates that neutral environmental cues can access amygdaloid circuitry of fear directly via thalamic connections, with no cortical intervention (Iwata et al. 1986) At present, the most fertile neuroscientific path to understanding emotions resides in the decoding of the subcortical circuitry which ultimately controls those intrinsic behavioral, autonomic, and feeling tendencies which characterize emotions. Until demonstrated otherwise, it is reasonable to assume that the affective states which characterize emotions arise from the same executive circuits which synchronize the other behavioral and physiological attributes of the primal emotions. It is to be anticipated that only at low levels of the neuroaxis, where the actual motor patterns are triggered, can affect and behavior become dissociated. While the role of the neocortex in generation of the affective content of emotions appears meagre, the role of subcortical emotional processes in coloring cortical functions with feelings is substantial.

There are many reasons to suppose that the manner in which the neocortex processes information may be dictated by the activities of subcortical emotional circuits. The referencing mechanisms for both the registration and retrieval of memories may be constituted of subcortical emotional circuits - one merely need consider energization of thinking and the profuse welling up of memories when emotions are provoked in humans. The work of Proust (e.g., Remembrance of Things Past) has become common currency for the literary highlighting of such affect-cognition interactions. Indeed, it is a common human experience that cognitions are guided in countless ways by our emotions.

Likewise, most learning tasks used in animal research have strong emotive components, and memories are laid down with especially high resolution in emotional contexts. Take just one excessively simple example, the "Coolidge effect" in sex research (Beamer et al. 1969) - males exhibit dramatically increased sex behavior when new females are made available, while simply disguising the "used" females does not restore sexual ardor. Obviously, the desire for sexual novelty can only occur through the auspices of primitive learning mechanisms. Indeed, every motivational process has powerful memorial processes associated with it: The Coolidge effect is as prominent in feeding research as in sex - give an animal new and tasty foods and it will gorge, compared to how it eats when given restricted access to a single item. Although we do not know the brain mechanisms by which such "simple" habituation-type memories are encoded, I suspect they highlight a key role of affective arousal in memory coding. Higher-order associative mechanisms may also be integrally linked to fluctuating activities of emotional systems (Blaney, 1986), with each emotive system having specific design characteristics for registration of pertinent memories (Panksepp 1986a).

The main obstacle to vigorous neurobiological research in the area of animal emotions is the logical difficulty of relating subjective emotional constructs to observable behavioral referents. Emotions are neither just sensory systems nor just motor systems, but ones which mediate, in preorganized ways, between those processes. The circuits are tuned to receive various sensory, perceptual and interoceptive stimuli to yield behaviorally flexible albeit functionally harmonious action patterns. In short, they are intervening executive systems which elaborate the instinctive nature of an animal in various life-challenging circumstances. Thus it is time for neurobiologically oriented investigators to assume that emotional behaviors are genuine and natural external signs for the activity of underlying emotional processes. To have thought otherwise (e.g. behaviorism) has proved to be a grave obstacle to the development of a credible science of behavior.

A HISTORICAL NOTE AND VIEW OF EMOTIONS VIA COMPUTER ANALOGY

For the better part of this century, experimental psychologists have sought to construct behavior out of simple sensory and motor processes, without resorting to constructs whereby behavioral orchestration is elaborated by intrinsic brain systems. Many past and present schisms in psychology can be better appreciated through a consideration of the multiple historical foundations for the analysis of psychological processes (see Fig. 1). The brain was long viewed by those who followed the philosophical tradition (arising initially from the writings of John Locke) as tissue which resembled random-access-memory (RAM) in modern computers, upon which the sensory-motor association networks of experience were written. This conception is simply wrong. Meanwhile, the analysis of brain functions which emerged from the biomedical tradition had a much more phrenological/localizationalistic bent, whereby parts of the brain were seen to be specialized for elaboration of specific functions. This conception was much closer to the reality of the

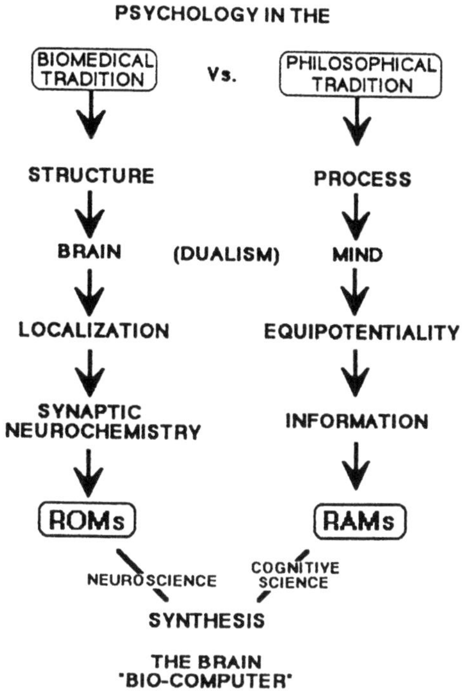

Fig. 1. The two major traditions by which psychological processes have been studied historically has yielded dualistic perspectives to our understanding of emotions and other brain functions. Both lines of thought should be blended into a synthesis which recognized that the brain-bio computer is constructed of both genetically-coded ROMs (read-only-memory) and experientially programmed RAMs (random-access-memory) functions. Biological understanding of emotions can be best facilitated if we pursue the psychological issues in humans and concurrently analyze the biological issues in the animal brain.

matter, but it is still not as influential as the more "democratic" notion which arose from philosophical roots.

While the localization perspective is now well accepted for sensory and motor processes, it is also now clear that many intervening subcortical behavior control circuits actually resemble read-only-memories (ROMs) which comprise operating systems for the generation of various types of ingrained psychobehavioral tendencies. It is presumably through the auspices of the genetically prewired subcortical ROMs, some of which elaborate primary-process emotionality, that many of the associative structures in neocortical RAM space are constructed. However, biological brains probably really do not contain any truly blank-slate type circuitry analogous to computer RAMs. Although the most complex multimodal association cortex may approximate such computational space, even those tissues are constrained by biological limits to the types of information they can organize. Most of the multimodal association areas of the human brain appear to be genetically predestined (or at least prepared) to subserve specialized psychological functions, including, perhaps, subtle emotional ones which characterize human nature (e.g., shame, envy, guilt, jealousy, a desire for revenge, romantic love, a sense of friendship, a occasional desire to help others). In addition, the experiential actions of subcortical ROM processes on the pseudo-RAM functions of higher association cortices may construct programmable read-only-memory (PROM) functions in higher areas which then dictate the emotional habits and personality traits of an individual. Indeed, many of the asymmetrical emotive functions which are being discovered in the cerebral hemispheres of humans may in fact be socially-constructed PROM functions which have arisen from the experiential, real-life interactions of perceptions with genetically-ordained subcortical emotional ROM functions. Thus, a clear analysis of the ROM functions may be essential for understanding the derivative functions.

Although the "behaviorist/radical empiricist" bias which sought to view the brain as a tabula rasa has diminished markedly in the past few decades, it has not yet abated to a point where certain obvious conceptual alternatives (e.g., ROM-like emotional circuits) are readily entertained. My personal research in the area is an attempt to see how far ordinary emotional concepts can take us toward an understanding of certain key aspects of brain organization. In this scheme, emotive command circuits are viewed as evolutionary inventions which help coordinate key behavioral tendencies (and their associated physiological and hormonal patterns). Labeling these circuits with common emotional labels is not an attempt to explain behavior, but rather represents the starting point of empirical investigations to understand the intrinsic neurobiological operating characteristics of the underlying brain circuits. The "hidden agenda" of this research is the hope that an understanding of such emotive operating systems will reveal how the brain elaborates adaptive and maladaptive behavior patterns in humans. To do this, neurological and psychological approaches must be merged.

As I have previously discussed in detail several times, there is presently convergent evidence for the existence of a small number of basic emotional command circuits which course through the diencephalon, linking higher limbic areas with lower brainstem and spinal reflex circuits (Panksepp 1981a, 1982, 1985a, 1985b). These systems include ones to elaborate fear, separation distress, anger, and exploration as well as play. By taking the perspective that any definition of emotions must include neural criteria (Panksepp 1982, 1986), we have been willing to view certain brain systems which have not traditionally fallen into the category as emotion-mediating ones. For instance, rather than viewing the self-stimulation system which courses through the lateral hypothalamus as either a reinforcement, reward, or pleasure system, I view it as one which mediates exploratory urges and the anticipatory excitement which energizes foraging behaviors and helps create positive expectancies about the external world. Ultimately, this system may be the driving force of human curiosity and the source of "psychic energy" which sustains interest-directed activities (Panksepp 1986). Although these words have many definitional ambiguities, I suspect that they can guide a more incisive analysis of the underlying brain systems and psychological processes than the stimululs-response/reinforcement conceptualizations of behaviorism.

The reason emotional terminology has not been widely used in animal research is because of the severe definitional problems which commonly emerge from the fact that all emotive terms ultimately refer to internal subjective states. One reasonable perspective on this problematic issue is that certain evident processes of the animal brain should initially be accepted as facts of nature which need to be biologically explored (with good operational definitions), rather than conceptually defined. If the brain contains various intrinsic control processes, how can we define them coherently before we understand something substantive about the nature of the underlying mechanisms? After all, the ultimate definition of a concept is the explication of the mechanism by which it is elaborated, and surely we cannot do the former credibly before we do the latter. The neural background processes of animal behavior need to be probed empirically before they can be defined in reasonably comprehensive ways. In the interim, we need to use words, and for those brain processes which I will consider herein, there are presently no better choices than the common emotional terms that have long been built into our languages. Although common words contain abundant ambiguities, and all too often idiosyncratic connotations for different individuals, the generation of a new lexicon would probably prove counterproductive - if for no other reason than the likelihood that a new set of terms will not be accepted by others, and may alienate young investigators from pursuing the critical neurobiological issues. For instance, during the past decade we have been analyzing social play behavior in rats (via behavioral anchors that are operationally well-defined), albeit we can provide no adequate conceptual definition of play (Panksepp et al. 1984, 1987). Still, the systematic nature of our data indicates that we are studying a harmonious brain process, which, for lack of any better terminology, is quite satisfactorily characterized as play. Although fellow investigators have expressed doubt over the validity of the concept (and we encountered difficulty in publishing our empirical findings for several years for such semantic reasons[1]), the true "experts" we have consulted, namely young children, have exhibited no problem with the concept. They recognize play when they see it, even though they can provide no better definition for it than "play is fun". We assume that when we have deciphered the subcortical brain circuitry for social play in rats, (since decortication leaves play essentially intact) we will have a basic idea of the brain mechanisms which generate the motivation for human social play. Likewise, we believe that when we understand the distress vocalization circuitry of the vertebrate brain, we

---

[1] Because of the resistance we have encountered in using the simple word play, we have entertained replacing it with the ludicrous acronym P.L.A.Y. (Prominent Ludic Activities of Youth). However it would be a delusion to believe that such illusory usages would help the scientific enterprise (please note that the root of the italicized words is the Latin ludare - to play).

are getting as close to the biological sources of human sorrow and grief as we presently can.

There is an empirically uncrossable, albeit not logically unbridgable, philosophical chasm between the subjective nature of human affect and the emotive behaviors of animals. Obviously the bridge must be theoretical. To make progress in this field, psychobiology has to accept that it is confronted by a quandry similar to that confronting particle physics (or for that matter the study of gravity), where the object of inquiry has to be observed indirectly. The present thesis is that the as if strategy can serve as a viable procedure by which to seek biological evidence concerning the basic nature of emotionality. In other words, if we assume that animals under specific circumstances are behaving as if their actions are being controlled by emotional processes in the brain, then we are in a position to begin unravelling some key issues concerning the sources of animal and human behavior.

The probability that this approach will yield credible scientific insights seems high, although it seems, to some, to lack the desired analytic rigor we expect from hard science. As I have summarized elsewhere, the as if approach yields some remarkable congruences between human emotional experience and animal emotive behaviors (Panksepp 1985). Although it remains improbable that we will ever be able to study directly the subjective experiences of other animals, I would suggest the as if approach can take us to the heart of those shared primitive processes which govern emotionality in all mammals. In short, the emotive behaviors of animals may be "natural metaphors" for the types of emotional feelings which exist inside their brains. Indeed, the importance of emotional communication among members of social species has guided the evolution of special-purpose mechanisms which allow animals to decode each other's feelings (e.g., the emotional flexibility of the human face, which other humans are perceptually prepared to decipher, apparently via right subtemporal lobe functions). Indeed, in humans, communicative skills have evolved to the point where external signs and the internal feelings can be dissociated (perhaps via the evolutionary value of deception in highly social creatures who must seek to get along). From this, one could argue that the study of basic emotions is more problematic in humans than in animals. Indeed, the differential neural controls for spontaneous and posed facial expressions affirms the emotional duplicity of humans (Kahn, 1964) and adds a troublesome dimension to the analysis of how the two hemispheres elaborate emotionality in humans.

Of course the as if approach, if used as more than a guide to "what to study," can yield very inadequate conclusions. It cannot be used as an explanation of behavior, only as a preliminary way of conceptualizing certain coherently operating brain systems that guide psychobehavioral processes. Accordingly, the subjective, affective experiences that most humans share with each other, without systematic past learning, is the most reasonable starting point for the postulation of the types of emotional systems that we ought to search for in the brain (Panksepp, 1982, 1985). Of course, subjective experience can only

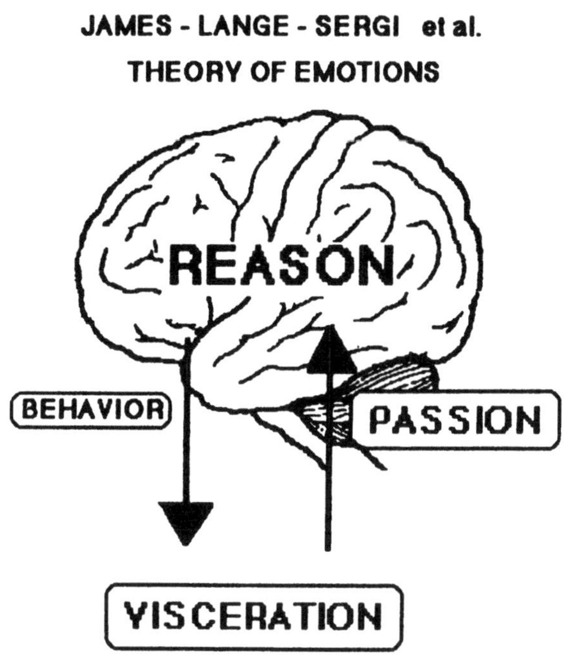

Fig. 2. The James-Lange perspective suggested that emotions arose from the tendency of a reasoning brain to evaluate the visceral commotion arising from vigorous motor activities. This viewpoint seems to be essentially incorrect.

identify what needs to be studied; it is a very poor guide to the nature of the specific mechanisms that actually exist in the brain (although it may provide clues as to the dynamics and interactions among mechanisms).

The above issues are highlighted by the historical evolution of thinking about emotions (Figs. 2,3). The classical perspective was that emotions were elaborated in the visceral organs of the body, while reason (perception,judgement,memory) was elaborated in the brain - a viewpoint which is certainly consistent with some primitive gut-feelings. This perspective was only marginally corrected by the James-Lange theory, which suggested that the reasoning brain reads peripheral, visceral, and somatic changes arising from vigorous behaviors as emotional states (Fig. 2). Although probably not completely incorrect, this perspective has little empirical support to commend it (e.g., Lowe and Caroll, 1985). In my opinion, the popularity of the James-Lange viewpoint has long hindered substantive progress in the field. However, it is understandable why so many human psychologists embraced the theory, for one can more readily measure peripheral autonomic changes in humans than the properties of brain circuits. The popularity of work on hemispheric control of emotions may have a comparable source - laterality correlates are more accessible than the subcortical circuits from which emotional feelings ultimately emerge. I would suggest, however, that a study of dorsal vs ventral (i.e., somatic vs visceral)

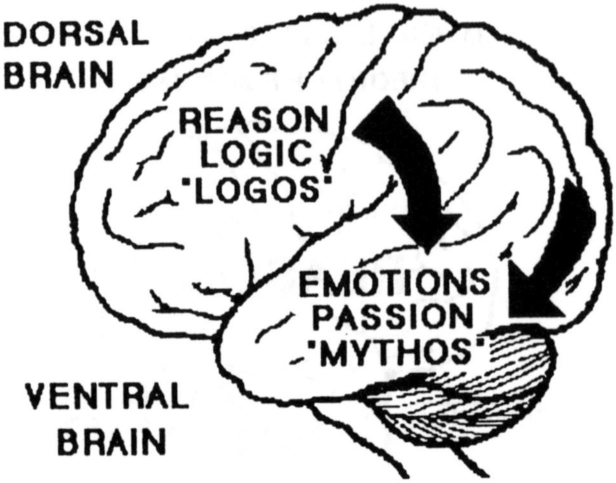

Fig. 3. The modern view is that emotionality is an intrinsic function of the ventral brain (i.e., limbic system), and that cognitive processes can activate emotive processes via gateways in specific areas of the brain, such as the temporal lobes

brain functions will be more important for out ultimate understanding of primary-process emotionality than a focus on right vs left brain issues (Figure 3).

It presently seems fairly certain that emotionality emerges from subcortical areas of the brain that have long been known to modulate visceral processes. The emotional vitality of the neodecorticate animal is perhaps the most compelling "proof". In view of such evidence, the James-Lange theory could have simply been revised to mean that emotions are cognitively appreciated by higher brain structures "viewing" the changing activities of these subcortical viscero/affective neural control systems which reside in subcortical areas of the brain. This is not to say that bodily states do not modify activities in emotional circuits. Surely, peripheral autonomic changes and fluctuating enteric chemistries exert some feedback control over the tone of the basic emotive circuits, albeit little is presently known about such reafferent mechanisms. Ultimately, these types of modulatory controls are conceptually quite different than those envisaged in the James-Lange theory. Also, it should be noted that many of the visceral feelings experienced during emotions may result directly from changing activities of the diffusely organized organ "homunculi" in the visceral nervous system as opposed to moment by moment afferent input from the viscera.

In fact, there is presently no credible evidence that specific emotions can be generated by experimentally induced changes in peripheral visceral tone. By comparison, it is remarkably easy to provoke emotions in humans and emotional behaviors in animals by directly stimulating the same specific areas of the visceral brain. As detailed elsewhere (Panksepp 1981a, 1982,

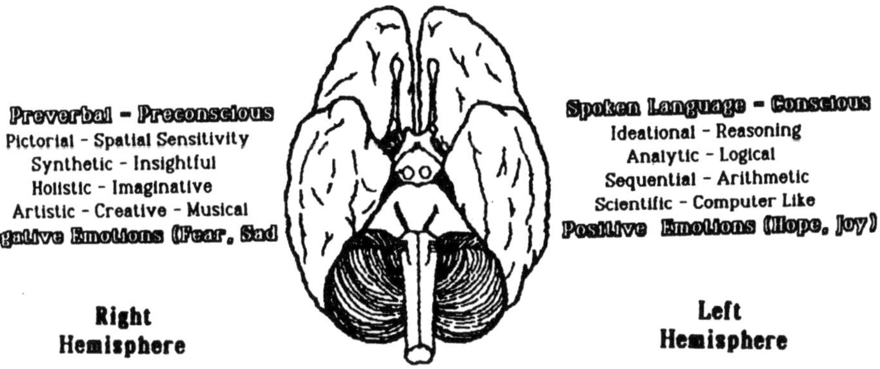

Fig. 4. An overview of hemispheric functions. It remains unclear why there might be a relationship between the language functions of the left hemisphere and positive emotionality. One possibility is that the right hemisphere became defensive due to the adaptive value of protecting the left side of the body (in which the heart was situated), leaving the right hand to be the more outwardly directed, world-manipulating limb. Since one major function of language is to represent the world that is being manipulated by the right hand, then the left hemisphere may have assumed a greater role in elaborating affective decisions related to foraging, expectancy, hunting, and other positive emotive states related to positive life-sustaining interactions with the outside world

1985, 1986a), brain stimulation studies provide reasonable evidence for the localization of fear-anxiety, separation distress-panic, rage-anger, and curiosity-expectancy systems of the brain; and other lines of evidence point to a play as a primary emotive system of the brain (Panksepp et al. 1984).

HEMISPHERIC CONTROL OF EMOTIONS

The present popularity of research into hemispheric specialization of emotion is based, at least in part, on the anatomical accessibility of this neuropsychological correlate. Although the apparent positive emotional bias of the left hemisphere and the negative emotional bias of the right hemisphere (Fig. 4) are reasonably robust and replicable phenomena (see Flor-Henry 1979; Ross 1984), there are quite a few inconsistent lines of evidence (Kolb and Milner, 1981; Strauss, 1986) and hence the specific meaning of the emotional laterality results remains ambiguous. The effects may reflect subcortical asymmetries in the organization of emotive processes, but there is little direct data to support that contention. Some recent evidence does indicate that dopaminergic systems which mediate schizophrenic symptoms may more strongly influence left temporal lobe processing (Reynolds, 1983). There is additional evidence for modest brain asymmetries in neurochemical systems, such as norepinephrine (Oke et al. 1978), serotonin (Mandell and Knapp, 1979) and acetylcholine (Gainotti et al. 1982). Hemispheric differences may also emerge from the differential access (perhaps learned) that each hemisphere has to subcortical emotive processes or to aysmmetrical hemispheric interactions. Finally, the cerebral asymmetry effects may reflect some innate aspect of emotional circuitry in each of the hemispheres. One attractive hypothesis along these lines is that each hemisphere has intrinsic abilities to organize certain perceptual aspects of the world which are designed to trigger emotional states. In other words, each hemisphere may be able to code certain environmental events into moods via specific multimodal association areas which can access subcortical emotive circuits. Although this comes very close to the postulation of prototypical innate ideas in the cortex, who can presently say that the complex human emotions of jealousy, shame, and guilt do not emerge from higher intrinsic potentials of the human brain? Certainly, such possiblities would coincide well with sociobiological conceptions of higher brain evolution, and it does suggest some new research strategies. In addition to formal psychometric testing, it may be useful for investigators to sensitively probe the changed dynamics of the real-life emotional stories and histories of their brain damaged patients.

We have devoted considerable effort to evaluating whether some of the basic emotive processes which are the focus of our laboratory work are differentially modulated by the two cerebral hemispheres. We have found no major effects in studies of (a) juvenile play in rats, (b) distress vocalizations in newborn chicks, or (c) courting calls in adult guinea pigs. Because of these negative results, the studies themselves remain unpublished, but the evidence has been as follows: (a) Total right or left neodecortication at 3 days of age left rough-and-tumble play of juvenile rats totally intact. Also, when right and left neodecorticates were pitted against each other, neither type of animal prevailed in terms of who pinned whom more. (b) When newborn chicks were unilaterally decerebrated down to the diencephalon, they responded to separation (in terms of frequency of distress vocalizations) similarly whether the right or left hemisphere had been aspirated. (c) Adult male guinea pigs purr rather predictably when "courting" a female, and the frequency of this response was normal in animals which had been deprived of either right or left neocortical mantles 2 weeks prior to testing. Whether right and left subcortical lesions have differential effects on emotional tendencies is largely unstudied. Although the animal work should be able to give clear answers as to whether subcortical emotive circuits are lateralized, there is presently very little work on the issue. At present, one of the few lines of behavioral evidence for asymmetries in subcortical emotive processes in animals consists of modest differences that have been observed in self-stimulation reward (Glick et al. 1980), but anatomical asymmetries such as those in cortical thickness (Diamond et al. 1981), transmitter levels (Camp et al. 1984; Dark et al. 1984; Denenberg and Rosen 1983), and the local organizational effects of gonadal steroids (Nordeen and Yahr 1982) suggest other emotive asymmetries may exist.

Still, our failure to find differential cortical modulation of several basic emotive tendencies makes us skeptical that the two hemispheres might differentially elaborate fundamental emotional tendencies, at least for the species and behaviors studied. Past demonstrations of emotional laterality in animals have typically utilized tasks which require substantial learning and/or past experiences (e.g., the recent laterality effects in rats demonstrated by Dennenberg (1981) and colleagues are all dependent on early infantile handling (Sherman et al. 1983). Perhaps the laterality that has been observed is related to differential emotional learning tendencies of the two hemispheres. This might be achieved by each hemisphere being differently attuned to symetrically organized emotive processes situated in subcortical areas. Obviously, resolution of such issues will require more research.

HIERARCHICAL CONTROL OF EMOTIONS

Although our own research and theoretical focus has been devoted to analyzing what we believe to be the fundamental subcortical circuits from which emotionality ultimately emerges, it seems evident that emotive systems are constituted of widely ramifying circuits that exert control throughout the neuroaxis. Accordingly, emotions have to be understood as multidimensional processes, and it seems inappropriate for any level of analysis, whether social, cognitive, or biologic, to claim that a unidimensional approach can capture all the facets of the phenomena subsumed by the concept of emotions. The new anatomy of the brain provides many pathways by which cognitive processes can come to modify limbic/emotive circuits, via temporal lobe, cingulate, and frontal cortices (Aggleton and Mishkin 1986; Cassell et al. 1986; Ervin and Martin 1986; Jurgens 1979; Kling 1986; Rolls 1986). Although such cortical influences on emotional circuits are becoming ever more evident, anatomically the influence of subcortical processes over cortical ones seems even more robust.

A most interesting neurological issue is whether higher reaches of the brain have specialized cortical areas for processing information and cognitions related to the various basic emotions. My reading of the literature is that there is adequate reason to believe that the basic emotive circuits which course through the diencephalon do have paleocortical spheres of influence, approximately as summarized in Fig. 5 - with expectancy processes being concentrated in basal forebrain and frontal cortical tissues, anger and fear in medial and lateral temporal lobe, respectively; and social emotions such as sorrow-grief-panic in cingulate areas. The literature on which these conclusions are based has been summarized elsewhere (Panksepp 1985) and will not be repeated here, except to note that the issue of intrinsic cerebral laterality of the basic emotions might be most fruitfully addressed by focusing on these areas of the brain

At lower levels of the neuroaxis, we will also probably find a variety of basic processes that modulate activity in emotive circuits. For instance, it is likely that metabolic and other homeostatic exchanges influence the tone of emotional circuits. Surely, hunger, thirst, and the status of various endocrine systems will modify activity in the foraging-expectancy, rage and play systems. The newly emerging neurochemistries of the enteric and immune

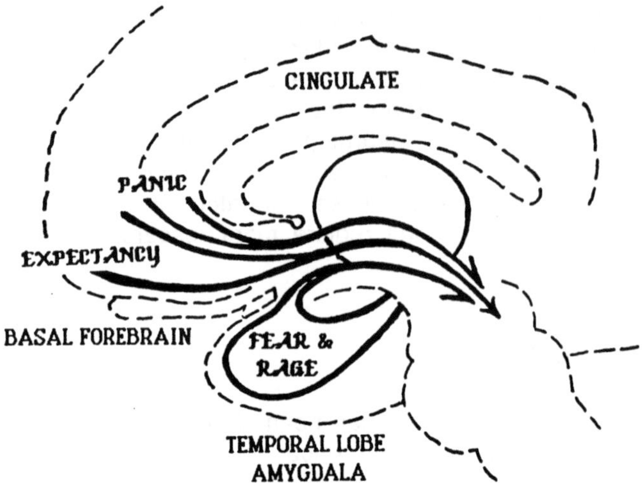

Fig. 5. Each emotive command circuit of subcortical areas of the brain probably has higher, telencephalic spheres of influence which were designed to elaborate cognitive/perceptual decision-making aspects for the fundamental emotions. The evidence behind the choices is more fully summarized in Panksepp (1985). It may be useful to focus on these areas of the telencephalon to clarify how the two hemispheres differentially control positive and negative emotions.

systems may well be found to influence the tone of emotive circuits. Such influences may be rather nonspecific, but two brain processes - those which subserve dreaming and playfulness - may be of special importance for articulating the other emotive functions of the brain, and may be especially important in laterality research.

THE ROLE OF DREAMING IN EMOTIONS

As observed in everyday experience, dreams are commonly laden with emotions. However, this observation has not been robustly affirmed by REM-awakening studies (Arkin et al. 1978; Foulkes 1982), with only 35% of REM dreams having clear emotional content (Snyder 1970). Nonetheless, there is still a high incidence of emotional dream content, especially if the aforementioned statistic does not include mild pleasant emotions related to positive incentives (perhaps reflective of the activity of expectancy and play systems of the brain). For instance, the REM dreams of very young children (3-5 years of age), although not abounding with negative emotions, do have abundant dream content related to "simple wants and pleasures" (p. 68, Foulkes 1982).

On the basis of work from Jouvet's laboratory we can be fairly confident that the REM dreams of other animals are quite laden with affect. The work of

(Sastre and Jouvet 1979), confirmed and extended by Morrison (1983), has found that animals with lesions of the dorsomedial tegmentum near the locus coeruleus do not exhibit the tonic muscular relaxation (i.e., motor neuron hyperpolarization) which normally characterizes REM sleep, and hence "act out" their dreams. During such "oneiric" activity, cats typically exhibit various emotive behaviors - predatory stalking, striking out in rage, fearful cowering, and grooming. Thus, one function of REM may be to access safely the major executive circuits for emotions, and thereby to facilitate the processing of information related to emotional events that occurred during waking (see Fig. 4.5, Panksepp 1986a).

Such a conception is congruent with the apparent information processing functions of REM. The most robust memories of an animal should be related to major survivial issues, which in waking were resolved through the auspices of emotional systems. If the emotional system permitted the generation of adaptive responses during waking, then a key function of REM may be to help integrate this new information into the permanent habit structure of the animal. From this perspective, the manner in which the REM process accesses the various emotive circuits (in terms of priorities, durations, and sequences) should become an important area of inquiry. Such data would be impossible to obtain were it not for the preparation developed by Sastre and Jouvet (1979). A statistical analysis of the "oneiric" behavior sequences exhibited by such animals could elucidate how REM mechanisms and emotive circuits interact. It is noteworthy that recent work suggests that posterior parts of the left hemisphere may be especially important in elaborating dreams (Greenberg and Caroll, 1987). It has also been known for some time that frontal lobectomies abolished dream content without much effect on the electrophysiological indices of REM activity (Jus et al. 1973).

EMOTIONS AND PLAY

Just as REM may help elaborate the affective perceptual content of emotive circuits during sleep, play may be specially designed to elaborate the behavioral possibilities inherent in emotive circuits during waking (Panksepp 1986a). As often noted, the sequences of behavior exhibited during social play have apparent emotive content without affective seriousness. Apparently the motor, as well as the sensory-affective side of emotional circuits, is being exercised. However, in such "joyful" ludic activities, animals take themselves to the very edge of their behavioral possibilities - to the edge of their existing worldviews, in a manner of speaking. It is at this borderland, where an animal's experiences are modest, that true emotional realities are most likely to intervene. At such points, play ceases and animals begin to respond "realistically" to the demands of their emotional circuits. In other words, a function of play may be to take an animal to the boundaries of its emotional knowledge, and thereby to become conversant with the edges of its psychological world. Of course such a pervasive brain system probably has multiple functions, including ones more basic than those outlined above. For instance, juvenile social play may help immunize an animal against future social stress during adult competition.

For such reasons, we have been devoting an ever-increasing amount of our research effort to understanding how play is organized in the brain. At this point, we are certain that there are special-purpose brain mechanisms designed specifically for the elaboration of play, and we have preliminary knowledge of the underlying neuroanatomies and neurochemistries, (Panksepp et al. 1984; Panksepp et al. 1987; Siviy and Panksepp 1985). We anticipate that a thorough neurobiological analysis of the underlying brain elaboration of play will demonstrate strong relations to executive neural systems for the basic emotions. Indeed, play circuitry may itself constitute a primal emotive system of the brain. If the left hemisphere of humans is truly more happy than the right, then it will be worth entertaining the possibility that it is influenced more by subcortical, playful tendencies (or arises from left hemispheric expressions of playfulness, as expressed in humans through verbal banter). Although we have not found evidence for left hemispheric ultra-playfulness in rats, those hemi-decortications were done in infancy, and it is possible that the asymmetric influences are PROM functions based on past experiences.

EMOTIONS AND PERSONALITY

It seems likely that personality structures are molded by the nature and vigor of emotional circuits possessed by an individual. It is commonly believed that the underlying structures of personality arise from the emotional dispositions (traits) of individuals. The classic temperaments (choleric, phlegmatic, melancholic and sanguine) are remarkably reminiscent of the basic emotive circuits of the brain (rage, fear, panic, and expectancy/play respectively). Since one of the functions of these systems is presumably the elaboration of habit structures, it should be of great interest to determine how emotional circuits govern the integrative activities of higher brain areas. There is already abundant evidence how various memories, especially early infantile and childhood learning experiences, strongly modify the structure of the brain. For instance, Spinelli et al.(1980), have demonstrated how avoidance learning with visual feedback, modifies the sensitivities and dendritic structure of somatosensory and visual cortices. Perhaps the observed brain changes were guided by the emotional circuits that were activated during the fear conditioning.

What happens to brain organization when an emotional system is repeatedly aroused during the early phases of development? Will persistent activation of the grief-panic system lead to a melancholic personality? Will chronic irritations and arousal of rage mold the child toward a choleric personality? Will repeated experience with fear lead to a phlegmatic temperament? Will the habitual arousal and reinforcement of the exploratory-foraging-expectancy and play systems lead to a sanguine personality? Will early emotional experiences permanently affect how adult brain circuitry and personality is organized? In conjunction with new neuroscience tools, existing animal models should be able to answer these questions, and the consequences for understanding the nature and construction of normal and abnormal personality structures may be enlightening. With such knowledge, we could rigorously evaluate the possibilities and limits of remedial change within the underlying psychoneural fabric, and also determine how emotional processing in the cerebral hemispheres becomes lateralized.

THE NEUROANATOMY/NEUROCHEMISTRY OF EMOTIONS

It would be an understatement to say that our understanding of emotional circuits is rudimentary. Many modern-day neuroscientists might question the claim that such circuits exist in the brain. There is no question that many of the arguments I have pursued are inferential, but a great deal of existing evidence affirms the existence of such neural systems. At the same time, there is not a single emotional circuit for which the command structure, either anatomically or neurochemically, has been decisively identified. Many of the neuropeptide pathways which have been recently revealed course through the diencephalon, connecting higher and lower areas of the neuroaxis, and these are excellent candidates for emotional "command-circuits" (Panksepp 1986c). For instance, atriopeptin-immunoreactive systems in the brain overlap with cardiovascular regulatory areas (Saper et al. 1985); growth hormone releasing factor (GHRF) administered into the brain yields prolonged bouts of aroused behavior suggestive of heightened emotionality of some type (Tannenbaum 1984); and corticotropin-releasing factor (CRF) administered intraventricularly yields similar forms of motor arousal (Britton et al. 1982). Since the central nucleus of the amygdala appears to be the very headwater for fear circuitry which descends via the anterior lateral hypothalamus to the central gray, the diversity of peptide systems concentrated in that nucleus is noteworthy (Cassell et al. 1986). Analysis of the specific emotive functions of peptide systems should be an exciting chapter of brain research during the next decade. In the following section, I will summarize evidence which suggests that the anatomy of the CRF system highlights the executive substructure of the emotional circuit which I have previously labelled the "panic" (or separation-distress) system.

THE NEURAL INFRASTRUCTURE OF SORROW, GRIEF, AND PANIC

Our work on separation distress (Panksepp et al. 1986) and other social emotions (Panksepp et al. 1984) is premised on the assumption that the findings will highlight the nature of human emotions related to social bonds and the severance thereof. Our work started with the theoretical assumption that there are physiological relationships between social dependence and narcotic dependence, and that opioid systems in the brain provided one powerful avenue of control over the negative emotions which emerge from lack of social support (Panksepp et al. 1980b; Panksepp 1981b). To model such processes in animals, we started to analyze separation induced "distress vocalizations" (DVs) and we soon learned that opioids, whether alkaloids or peptides, are uniquely efficacious in reducing separation distress (for review see Panksepp et al. 1986b). Although most other psychoactive agents had relatively little effect, some agents, which have important implications for addictive behavior, like nicotine (Sahley et al. 1981) and clonidine (Rossi et al. 1983), could yield potent reductions of DVs. Indeed, although several other pharmacological agents were found to increase separation-induced DVs - such as serotonergic, opioid, and muscarinic acetylcholine receptor blockade - none sustained distress when animals were returned to their companions (Panksepp et al. 1980a). At the present, the only agents that we have found to yield persistent crying in chicks which could not be quelled by companions is that induced by ventricular administration of curare (Panksepp et al. 1984), kainic acid (KA) (Normansell et al. 1987), and CRF (Panksepp et al. 1987). Under all of these conditions, chicks exhibit markedly increased separation induced DVs when isolated, and at high doses, they continue to vocalize even when returned to the flock.

The effects of the first two of these agents, curare and kainic acid, are hard to interpret, not only neurochemically but behaviorally. First, using traditional pharmacological drug-interaction strategies, we have not been able to clearly specify the receptors via which curare and kainic acid operate. Behaviorally, the animals appear to be highly agitated and exhibit not only an increase in vocalizations, but intense fear responses. Also, their visual capacities are compromised. Accordingly, it is possible that these agents are affecting several distinct forms of trepidation, as well as non-affective sensorimotor functions. The effects with CRF, though, have been behaviorally cleaner.

The reason we started to study CRF was based exclusively on the striking anatomy of the neural circuit trajectories as revealed in the rat brain. The cell fields, as well as the axonal trajectories (Swanson et al. 1983), course through areas of the brain which we had already revealed as those sustaining separation distress type vocalizations in response to localized electrical stimulation (Herman and Panksepp 1981). Areas that are especially effective in precipitating DVs are the ventral septum, bed nucleus of the stria terminalis, dorsomedial thalamus, and areas in and adjacent to the dorsolateral periventricular gray - all areas which "light up" in CRF immunocytochemical studies.

We have completed a thorough analysis of intraventricular CRF on the DV response of domestic chicks (Panksepp et al. 1987). The results were clear-cut. CRF at doses above 0.2 μg eliminated the tendency of young chicks to reduce their output of DVs in the presence of mirrors. At doses higher than 1 μg, the animals would continue to vocalize when returned to the flock, but unlike curare and KA-treated chicks, there was no accompanying intense fear/flight pattern during which animals often segregated themselves from the rest of the flock. In fact, CRF-treated animals very clearly appeared to congregate more and they seemed to be ducking under other animals and typically appeared as if they were vying for a position in the center of the flock.

The results are congruent with the possibility that the CRF system is the primitive executive substrate for separation-distress, and hence by inference, a major biological source of human emotions which arise from social separation (e.g., loneliness, sorrow, and grief). If this is the case, the predictions are clear. First, future immunocytochemical work on the CRF system in other species should clearly outline vocalization circuitry as revealed by other procedures (e.g., electrical stimulation and lesions). Indeed, recent anatomical mapping of DV circuitry in birds suggests strong neural connections to brain areas where CRF cells are typically found in mammals (Akesson et al. 1987; Cheng et al. 1987). It might also be predicted that centrally applied CRF receptor blocking agents and CRF antibodies should alleviate natural separation distress. These tests remain to be done, but to this investigator's knowledge, the CRF system presently stands as a preeminent candidate for an executive circuit for emotionality that has emerged from neuroscience research. But there are other promising systems on the horizon.

Although the spin-offs from such research to clinical practice will surely be slow to emerge, the ability of such work to promote new and effective therapies in psychiatric practice may be the most important criterion of the validity of the underlying theoretical structures. Already there is a congruence between studies of drugs which modulate separation distress and clinical effects on humans experiencing undesirable emotional responses related to separation. For instance, the tricyclic antidepressants are effective in alleviating separation distress in animals (Scott 1974; Suomi et al. 1978), as well as panic attacks and other separation syndromes such as school phobias (Klein 1981). They should be effective in reducing human sadness and crying.

Our work on brain opioids and social emotions led to the idea that those systems may be overactive in the brains of autistic children (Panksepp 1979; Panksepp and Sahley 1987), and preliminary clinical trials (Campbell et al. 1985; Herman et al. 1986) suggest that opioid receptor blockade with naltrexone can alleviate various symptoms of the congenital CNS disorder. To quote Campbell: Three (of five) "children were responders; therapeutic effects were tranquilizing (decreases of hyperactivity, aggressiveness, negativism and irritability) and stimulating (decreases of withdrawal, increases of eye contact, social overtures, language production and attention span)." Such correspondences speak of a fruitful rapproachment between the basic analysis of emotional circuits in the animal brain and the search for therapeutic maneuvers in the human clinic.

IN SUMMARY

Considering the evident fact that the relatively recent brain evolution in mammalian species has mushroomed in neocortical areas that are designed to deal with information arising from the external world, in conjunction with the evidence that the impulse for behaving emotionally emerges from more primitive subcortical structures that are still shared quite equally by all mammals, a reasonable conclusion is that we should be able to study the sources of human emotionality, albeit perhaps not the sources of human reason, in the animal brain.

Although there are some suggestive findings regarding the laterality of various basic processes in the animal brain (e.g., Denenberg 1981), the source of emotional specialization of the two sides of the human brain may be in the way the hemispheres deal with the bilaterally represented information arising from lower areas in the brain. This specialization could take the form of differential emotional learning about and differential inhibition of subcortical circuits. In short, asymmetries in emotional control may emerge from the selective ways in which relatively symmetrical subcortical influences are processed and elaborated by the two hemispheres.

Alternatively, the two hemispheres may have separate intrinsic functions in processing emotional information, but this may only exist in highly cerebrated species which actively use emotional information for communication. Although the data on human facial asymmetries are by no means unambiguous (Silberman and Weingartner, 1986; Strauss, 1986) the left hemisphere may have specialized functions for the voluntary/posed (at times deceptive) transmission of facial information realted to emotions (Cacioppo and Petty, 1981), especially in females trying to express positive emotions (Borod et al 1986), while the right hemisphere may have specialized functions for the reception and affective decoding of such information (e.g., Borod et al. 1986; Kolb et al 1983). Since it seems quite clear that the right hemisphere is more adept at decoding incoming emotional information, one can envision there having been an evolutionary pressure for the right side of the face to develop the voluntary communicative skills to take advantage of the left visual fields of others. The fact that spontaneous emotionality is expressed more intensely on the left side of the face may thus simply be a residual consequence, since the right side of the brain is normally not as adept at voluntary manipulation of the facial musculature.

If the above scenario is accurate, research on the brains of common laboratory animals, which do not voluntarily use emotions for communicative purposes, but use emotions only for primitive instrumental ends, may provide little insight into differential hemispheric elaboration of emotional communication in humans. Whether there is some type of preadaptation in the animal brain which guided the eventual specialization of the the human hemispheres is probably a distinct issue which still needs further exploration. In any event, at present it seems that animal research is more useful for unravelling the subcortical control of hormonal, physiological and primal affective/behavioral attributes of emotions that all mammals currently share. At present, these primal systems appear to be organized quite symmetrically below the cortex of the mammalian brain. How they link up with higher cortical functions remains largely unfathomed, but a key area for future research. When that level of knowledge is better developed, we may be able to discern why affective processing of the higher reaches of the human brain took the paths they did. At present we do not know (for one scenario, see the Legend of Fig 4). In any event, as we gain more information about the underlying neural systems of the basic emotions, what previously appeared complicated and capricious, may come to be seen as essentially simple, and in a deep sense, orderly.

REFERENCES

Aggleton JP, Mishkin M (1986) The amygdala: sensory gateway to the emotions. In: Plutchik R, Kellerman H (eds) Emotion: theory, research and experience, vol 3, Biological foundations of emotion. Academic, New York, pp 281-299

Akesson TR, de Lanerolle NC, Cheng MF (1987) Ascending vocalization pathways in the female ring dove: projections of the nucleus intercollicularis. Exp Neruol 95: 34-43

Arkin AM, Antrobus JS, Ellman SJ (eds) (1978) The mind in sleep: psychology and psychophysiology. Erlbaum, Hillsdale, NJ

Beamer W, Berman G, Clegg M (1969) Copulatory behavior of the ram, Ovis aries. II. Factors affecting copulatory satiety. Animal Behav 17: 706-711

Blaney PH (1986) Affect and memory: A review. Psych Bull 99: 229-246.

Borod, JC, Koff E, Buck R (1986) The neuropsychology of facial expression: data from normal and brain-damaged adults. In: Black D, Buck R, Rosenthal R (eds) Nonverbal Communication in the Clinical Context. Pennsylvania State Univ. Press, University Park, pp 196-222

Britton DR, Koob GF, Rivier J, Vale W (1982) Intraventricular corticotropin-releasing factor enhances behavioral effects of novelty. Life Sci 31: 133-139

Cacioppo JT, Petty RE (1981) Lateral symmetry and expression of cognition and emotion. J exp Psych 7: 333-341

Camp DM, Robinson TE, Becker JB (1984) Sex differences in the effects of early experience on the development of behavioral and brain asymmetries in rats. Physiol Behav 33: 433-439

Campbell M, Small AM, Perry R, Palij M, Nobler M, Polonsky B, Shore H (1985) Naltrexone in autistic children. In: 17th annual meeting and conference of the National Society of Children and Adults with Autism, Autism Research Symposium, Los Angeles, CA

Cassell MD, Gray TS, Kiss JZ (1986) Neuronal architecture in the rat central nucleus of the amygdala: A cytological, hodological and immunocytochemical study. J comp Neurol 246: 478-498

Cheng, MF, Akesson, TR, de Lanerolle, NC (1987) Retrograde HRP demonstration of afferent projections to the midbrain and nest calls in the ring dove. Brain Res Bull 18: 45-48
Dark, KA, Ellman, G, Peeke, HVS, Galin D, Reus VI (1984) Sex differences and symmetries of cateholamines: Relation to turning preferences. Pharmacol Biochem Behav 20: 327-330.
Denenberg VH (1981) Hemispheric laterality in animals and the effects of early experience. Behav Brain Sci 4: 1-49
Denenberg, VH, Rosen GD (1983) Interhemispheric coupling coefficents: Sex differences in brain neurochemistry. Am J Physiol 24: R151-R153
Diamond MC, Dowling GA, Johnson RE (1981) Morpholoigc cerebral cortical asymmetry in male and female rats. Exp Neurol 71: 261-268
Erwin FR, Martin J (1986) Neurophysioloigcal bases of the primary emotions. In: Plutchik R, Kellerman, H (eds) Emotion: theory, research, and experience, vol. 3, Biological foundations of emotion. Academic, New York, pp145-170
Flor-Henry P (1979) On certain aspects of the localization of the cerebral systems regulating and determining emotions. Biol Psychiatry 14: 677-698
Foulkes D (1982) Children's dreams: longitudinal studies. Wiley, New York
Gainotti G, Sorbi S, Miceli M, Amaducci L (1982) Neuroanatomical and neurochemical correlates of cerebral dominance: A mini review. Res Comm Psychol Psychiat Behav 7: 7-19
Glick SD, Weaver LM, Meibuch RC (1980) Lateralization of reward in rats: Differences in reinforcing thresholds. Science 207: 1093-1095
Greenberg MS, Farah MJ (1986) The laterality of dreaming. Brain Cognit 5: 307-321
Herman BH, Panksepp J (1981) Ascending endorphin inhibition of distress vocalization. Science 211: 1060-1062
Herman BH, Hammock MK, Arthur-Smith A, Egan J, Chatoor I, Zelnik N, Corradine M, Appelgate K, Boeckx RL, Sharp SD (1986) Role of opioid peptides in autism: effects of acute administration of naltrexone. Neurosci Abstr 12: 1172
Iwata J, LeDoux JE, Meeley MP, Arneric S, Reis DJ (1986) Intrinsic neurons in the amygdaloid field projected to by the medial geniculate body mediate emotional responses conditioned to acoustic stimuli. Brain Res 383: 195-214
Jurgens U (1979) Neural control of vocalization in nonhuman primates. In: Steklis HD, Raleigh MJ (eds) Neurobiology of social communication in primates. Academic, New York
Jus A, Jus K, Villeneuve, A, Pires A, Rachance R, Fortier, J, Villeneuve R. (1973) Studies on dream recall in chronic schizophrenic patients after prefrontal lobotomy. Biol Psychiat 6: 275-293
Kahn EA (1964) Facial expression. Clin Neurosurg 12: 9-22
Klein DF (1981) Anxiety reconceptualized. In: Klein DF, Rabkin J (eds) Anxiety: new research and changing concepts. Raven, New York, pp 235-263
Kling AS (1986) The anatomy of aggression and affiliation. In: Plutchik R, Kellerman H (eds), Emotion: theory, research, and experience, vol. 3, Biological foundations of emotion. Academic, New York, pp 281-299

Kolb B, Milner B (1981) Observations on spontaneous facial expression after focal cerebral excisions and after intracrotid injections of sodium amytal. Neuropsychologia 19: 505-514

Kolb B, Milner B, Taylor L (1983) Perception of faces by patients with localized cortical excisions. Can J Psychol 37: 8-18

Lowe J, Carroll D (1985) The effects of spinal injury on the intensity of emotional experience. Br J Clin Psychol 24: 135-136.

Mandell AJ, Knapp S (1979) Asymmetry and mood, emergent properties of serotonin regulation Arch Gen Psychiat 36: 909-916

Morrison AR (1983) A window on the sleeping brain. Sci Am 248: 94-102

Nordeen EJ, Yahr P (1982) Hemispheric asymmetries in the behavioral and hormonal effects of sexually differentiating mammalian brain. Science 218: 391-393

Normansell L, Zeisloft D, Panksepp J (1987) Effect of glutamate agonists and antagonists on calling in domestic chicks. Neurosci Abstr 13: 763

Oke, A, Keller R, Medford I, Adams RV (1978) Lateralization of norepinephrine in human thalamus. Science 200: 1141-1413

Panksepp J (1979) A neurochemical theory of autism. Trends Neurosci 2: 174-177

Panksepp J (1981a) Hypothalamic integration of behavior: rewards, punishments and related psychological processes. In: Morgane PJ, Panksepp J (eds) Handbook of the hypothalamus. Dekker, New York, pp 289-431

Panksepp J (1981b) Brain opioids: a neurochemical substrate for narcotic and social dependence. In: Cooper SJ (ed) Theory in psychopharmacology, vol 1. Academic, London, pp 149-17

Panksepp J (1982) Toward a general psychobiological theory of emotions. Behav Brain Sci 5: 407-468

Panksepp J (1985) Mood changes. In: Vinken PJ, Bruyn GW, Klawans HL (eds) Handbook of clinical neurology, revised series, vol 1(45). Elsevier, Amsterdam, pp 272-285

Panksepp J (1986a) The anatomy of emotions. In: Plutchik R, Kellerman H (eds) Emotions: theory, research and experience, vol 3, Biological foundations of emotions. Academic, New York, pp 91-124

Panksepp J (1986b) The psychobiology of prosocial behaviors: separation distress, play and altruism. In: Zahn-Waxler C, Cummings EM, Iannotti R (eds) Altruism and aggression: biological and social origins. Cambridge Univ Press, Cambridge, pp 19-57

Panksepp J (1986c) The neurochemistry of behavior. Ann Rev Psychol 37: 77-107

Panksepp J, Abbott B, Crepeau L, Sacks D (1989) The neuro- and psychobiology of fear systems in the brain. In: Denny MR (ed) Aversive events and behavior. Erlbaum, Hillsdale NJ, In Press

Panksepp J, Bean NJ, Bishop P, Vilberg T, Sahely TL (1980a) Opioid blockade and social comfort in chicks. Pharmacol Biochem Behav 13: 673-683

Panksepp J, Crepeau L, Clynes M (1987) Effects of CRF on separation distress and juvenile play. Neurosci Abstr 13: 1320.

Panksepp J, Herman BH, Vilberg T, Bishop P, DeEskinazi FG (1980b) Endogenous opioids and social behavior. Neurosci Biobehav Rev 4: 473-487

Panksepp J, Normansell L, Cox JF, Crepeau LJ, Sacks DS (1987) Psychopharmacology of social play. In: Mos J (ed) Ethopharmacology of social behavior. Nijhoff, Dordrecht, Netherlands

Panksepp J, Normansell L, Siviy S, Buchanan A, Zolovick A, Rossi J, Conner R (1983) A cholinergic command circuit for separation distress? Neurosci Abstr 9: 979

Panksepp J, Sahley T (1987) Possible brain opioid involvement in disrupted social intent and language development in autism. In: Schopler E, Mesibov G (eds) Neurobiological issues in autism. Plenum, New York.

Panksepp J, Siviy S, Normansell L (1984) The psychobiology of play: theoretical and methodological perspectives. Neurosci Biobehav Rev 8: 465-492

Panksepp J, Siviy S, Normansell L (1985) Brain opioids and social emotions. In: Reite M, Fields T (eds) The psychobiology of attachment and separation. Academic, New York, pp 3-49

Reynolds GP (1983) Increased concentrations and lateral asymmetry of amygdala dopamine in schizophrenia. Nature 305: 527-529

Rolls ET (1986) Neural systems involved in emotions in primates. In: Plutchik R, Kellerman H (eds) Emotion: theory, research, and experience, vol 3, Biological foundations of emotion. Academic, New York, pp 281-299

Ross ED (1984) Right hemisphere's role in language, affective behavior and emotion. Trends Neurosci 7: 342-346

Rossi J, Sahley TL, Panksepp J (1983) The role of brain norepinephrine in clonidine suppression of isolation-induced distress in the domestic chick. Psychopharmacol (Berlin) 79: 338-342

Sahley TL, Panksepp J, Zolovick AJ (1981) Cholinergic modulation of separation distress in the domestic chick. Eur J Pharmacol 72: 261-264

Saper CB, Standaert DG, Currie MG, Schwartz D, Geller DM, Needleman P (1985) Atriopeptin-immunoreactive neurons in the brain: presence in cardiovascular regulatory areas. Science 227: 1047-1049

Sastre PJ, Jouvet M (1979) Le comportement onirique du chat. Physiol Behav 22: 979-989

Scott JP (1974) Effects of psychotropic drugs on separation distress in dogs. Proceedings IX Congress Neuropsychopharmacology (Paris). Excerpta Medica Amsterdam, pp 734-745 (International Congress series no 359)

Sherman GF, Garbanati JA, Rosen GC, Hofmann M, Yutzey DA, Denenberg VH (1983) Lateralization of spatial preference in the female rat. Life Sci 33: 189-193.

Silberman EK, Weingartner H (1986) Hemispheric lateralization of functions related to emotion. Brain Cognit 5: 322-353.

Siviy SM, Panksepp J (1985) Dorsomedial diencephalic involvement in the juvenile play of rat. Behav Neurosci 99: 1103-1113

Snyder F (1970) The phenomenology of dreaming. In: Madow L, Snow LH (eds) The psychodynamic implications of the physioloigcal studies on dreams. Thomas, Springfield, Ill, pp 124-151

Spinelli DH, Jensen FE, DiPrisco GV (1980) Early experience effect on dendritic branching in normally reared kittens. Exp Neurol 62: 1-11

Suomi SJ, Seaman SF, Lewis JK, DeLizio RD, McKinney WT Jr (1978) Effects of imipramine treatment on separation induced disorders in rhesus monkeys. Arch Gen Psychiat 35: 321-325

Swanson LW, Sawchenko PE, Rivier J, Vale WW (1983) Organization of ovine corticotropin-releasing factor immunoreactive cells and fibers in the rat brain: an immunohistochemical study. Neuroendocrinol 36:165-186

Tannenbaum GS (1984) Growth hormone-releasing factor: direct effects on growth hormone, glucose, and behavior via the brain. Science 226: 1270-1277

Whishaw IQ, Kolb B (1984) Behavioral and anatomical studies of rats with complete or partial decortication in infancy. In: Almli CR, Finger S (eds) Early brain damage, vol 2, Neurobiology and behavior. Academic, New York, pp 117-138

Some Anatomical Substrates of Emotion, and their Bihemispheric Coordination

R. W. Doty

Department of Physiology, University of Rochester Medical Center, Rochester, New York 14642, USA

INTRODUCTION

Reproductive success is the engine of evolution. In the life cycle of vertebrates such success obviously requires an extended survival of individuals, as well as assurance of their fecundity by appropriate sexual and maternal or paternal behavior. Since these activities, survival and sex, are thus the sine qua non of vertebrate life, they have been primary forces in the evolution of the nervous system. Indeed, for non mammalian vertebrates, almost all neuronal activity can be placed within this framework; and while the elaboration of neocortex has given mammals a vastly augmented repertoire of formulas for personal survival, these cortical systems remain under preemptive control from, and interact continuously with, the more primitive and fundamental neural components with which they have evolved.

In regard to individual survival, homeostasis is the key factor, maintenance of appropriate body temperature, nutrition, and avoidance of pain. Emotional tone reflects, in discomfort or contentment, the success or failure of this homeostatic achievement. The achievement and its associated emotions, however, can be extremely complex, as in predatory stalking and attack, or the constant fearful alerting of the potential victim, which in the scheme of things often includes the predator itself. In such a threatening environment, in which must also be considered the pressures of territoriality, the interindividual contact (see Schloeth 1956; Leyhausen 1973) needed for reproduction requires some very special arrangements which, in turn, have their powerful emotional concomitants. Intraspecific aggression among males is a primary feature of reproductive strategy, as is a variety of olfactory and vocal cues in both sexes, tied to and evoking vivid emotional expression. The higher primates are to a large degree unique in also having integrated these basic emotional patterns into a societal life style, where interdependency among individuals in the group is critical for their for long-term survival (Carpenter 1964; Hall and Devore 1965; Bertrand 1969; Napier and Napier 1985; Goodall 1987). Status within the group is subject to continual adjustment and, as evidenced by the human participants in such societal arrangements, is commonly accompanied by significant emotional response.

Experimental Brain Research Series 18

It seems probable that it is the interplay between neocortex and the subcortical core of survival and sexual circuitry which produces the subjective state of emotion in ourselves. That statement, however, is largely supposition, since (a) it is difficult to obtain objective measures of such subjective states (see Panksepp 1982; Frijda 1986; Ledoux 1987), and (b) human cases with focal, bilateral loss of subcortical areas presumably relevant to emotional experience are exceedingly rare or nonexistent, so that it remains uncertain as to whether neocortical activity per se can give rise to emotional experience. It might be supposed that the more subtle, learned, aesthetic emotions could arise from neocortical activity alone, but the question is unlikely to be answered soon. In any event, the subtleties of social interaction, particularly those in higher primates, demand an exceedingly high level of visual analysis and mnemonic ability. Thus, any idea of emotion in an intact mammal being played out purely via subcortical circuitry is an unsustainable abstraction.

On the other hand, the evidence is unequivocal that subcortical structures are essential for the expression of the more "primitive" emotions, and can support such expression in the absence of neocortex.

Some Evidence for Subcortical Organization of Emotion

One of the classical experiments in the study of emotional behavior is that of Rioch and Brenner (1938), in which they showed that the "basic personality" of domestic cats is independent of neocortex. They selected five cats that were gentle and friendly, and two that were vicious and resisted handling of any kind. The neocortex was then removed from all seven animals, leaving the amygdala, olfactory tubercle, basal ganglia and most of the hippocampus intact. Following an extended period of recovery, the gentle cats still purred and rubbed when petted, but the vicious cats were unchanged, growling whenever touched. The gentle cats could, however, also be provoked into defensive reactions by painful stimuli. The converse experiment, performed by Bard and Rioch (1937), showed profound perturbation of preoperative behavior patterns when subcortical components were extirpated along with neocortex. Comparable observations have now been made on the opposum (Hara and Myers 1973) and hamsters (Murphy et al. 1981).

The differences in the basic behavioral disposition of various individuals (e.g., between newborn chinese versus caucasians! Freedman and Freedman 1969) is probably reflected in the anatomy and neurochemistry of components of the limbic system, especially between closely related wild versus domesticated animal strains; but such analyses are still preliminary (see Tizabi et al. 1979; Crusio and Schwegler 1987; Driscoll et al. 1987). Analysis will be complicated by the fact that various inherent characteristics are modifiable by experience (Valenstein et al. 1970; Karli et al. 1972; Adamec et al. 1980); even to the point that kittens at weaning learn to select preferentially such inappropriate food as bananas in imitation of a mother induced to do so by electrical stimulation of the hypothalamus (Wyrwicka 1978).

The locus of at least some of these elementary "emotional" traits is below the forebrain. For instance, it is clear for both man and rat that the facial expressions in response to taste stimuli, which in adult human subjects are accompanied by uniformly specific hedonic evaluation, are fully developed in the decerebrate condition (Grill and Norgren 1978; Steiner 1979). Of particular interest in the present context is the claimed hemispheric asymmetry in the eeg evoked by these stimuli in human neonates (Fox and Davidson 1986).

There is considerable evidence to suggest that the fore brain in many instances exerts a continuous, tonic, inhibitory control over these mesencephalic "centers" (see Roberts 1974; Doty 1976), and that their activation demands specific disinhibition from the forebrain. Such an effect is clearly seen, for instance, with grooming activity, which is more effective and more readily elicited in decerebrate rats than in those with the diencephalon is also present (Grill and Norgren 1978). A similar "release"probably accounts for the heightened attack behavior in decerebrate kittens (Bignall and Schramm 1974), or even the pathological laughing and/or crying occasionally seen in human patients following large cortical lesions or atrophy (Poeck 1969). It is of great significance that in many cases of pathological crying or laughing these extreme motor manifestations are devoid of emotional concomitant for the patient. However, this interesting dissociation between overt expression and subjective experience is not uniformly present; and electrical stimulation of these midbrain mechanisms in alert patients can arouse intense feelings of fear (Nashold et al. 1969). Such stimulation in anesthetized cats, squirrel monkeys or macaques can produce a dramatic expression of aggressive behavior, especially vocalization; and in the cat including piloerection, flattening of the ears, and striking with bared claws (Doty unpublished; see Clemente and Chase 1973; Adams1979 for review). Bandler et al. (1985) have still more precisely delimited the origin of such effects by showing that they can be elicited by excitatory amino acids injected into the periaqueductal gray, but not if injected into the surrounding tegmentum.

The data from both animal and human decerebration, and that from electrical and chemical excitation of the midbrain, are thus concordant in suggesting that neural systems in or near the periaqueductal gray integrate, or have access to, a variety of facio-vocal and other actions concomitant to at least the vigorous expression of aggressive and/or defensive emotion.

THE LIMBIC SYSTEM

This year marks the centennial of Brown and Schfer's observation (1888) that bilateral loss of the temporal lobes induces a remarkable tameness in otherwise wild macaques. It was another 50 years before Klüver and Bucy repeated and significantly extended such studies, revealing the dramatic changes in emotional responsivity consequent to complete loss of rostral temporal neocortex, amygdala, hippocampus and hippocampal gyrus (Bucy and

Klüver 1955). Concurrently, in a masterful synthesis, Papez (1937) proposed that emotional experience arises from an intricately connected neural system comprising the hippocampus, mamillary bodies, anterior thalamic nuclei and cingulate cortex. It remained, then, for MacLean (e.g., 1952), using anatomical, physiological and behavioral data from man and animals, to coin the term and concept of "the limbic system", emphasizing its relatedness to autonomic functions and emotional expression. The term has never been precise, particularly for its subcortical components, which were soon extrapolated into the midbrain. "Limbic system" largely subsumed and replaced the idea of a "rhinencephalon" (Jacobs et al. 1971). However, the phylogenetic origin of the limbic system in olfaction is worthy of emphasis; for in the depths of the sea or the dark of night olfaction provides perhaps the only cue to approaching danger, to a meal, or to a mate, and consequently to the related emotions and their expression.

Another point to be remarked is that although the term "limbic system" is often viewed as being something of an overly enthusiastic extrapolation, a monoclonal antibody has been developed (Levitt 1984) which has a high affinity almost exclusively for elements, both cortical and subcortical, commonly considered to be part of the system.

For discussion, but certainly not for function, the system can be conveniently divided into four major components: (1) the limbic cortex per se, cingulate gyrus and retrosplenial area (Yakovlev and Locke 1961; Baleydier and Mauguiere 1980), the allocortical region of the medial temporal lobe, i.e., parahippocampal gyrus and entorhinal area, and the orbitofrontal cortex, which were the major components of Maclean's original definition (1952). These neo and allocortical areas are interconnected by the cingulum and uncinate fasciculus, running as a great loop around the limbus of the corpus callosum. (2) The hippocampal system to which, of course, the entorhinal cortex and parahippocampal gyrus also belong, but which includes as well subiculum, dentate gyrus, cornu Ammonis, fornix, septal nuclei, mamillary bodies, mamillothalamic tract, anterior thalamic nuclei, and possibly nucleus reuniens (Rosene and van Hoesen 1987). (3) The many nuclei of the amygdala and their prolific connections with the hypothalamus. (4) The habenulo-interpenduncular complex, with its connecting tracts, the stria medullaris and fasciculus retroflexus (habenulo-peduncular tract) and numerous extensions into the midbrain tegmentum.

In accord with the view expressed above, that subcortical circuits are essential for emotional expression, and probably for experience as well, the neocortical components of the limbic system, 1) above, will not receive further attention herein. Similarly, discussion of the hippocampal system will be omitted, since it is now clear that its primary role is likely to lie in the mnemonic recording or retrieval of events (Olton et al. 1979; Zolamorgan et al. 1986; Muller et al. 1987). While emotion may indeed be related to events, for most such records it is probably a minor factor. On the other hand, the amygdala unquestionably, and probably the habenulo-interpenduncular complex as well, play a major role in emotional behaviors.

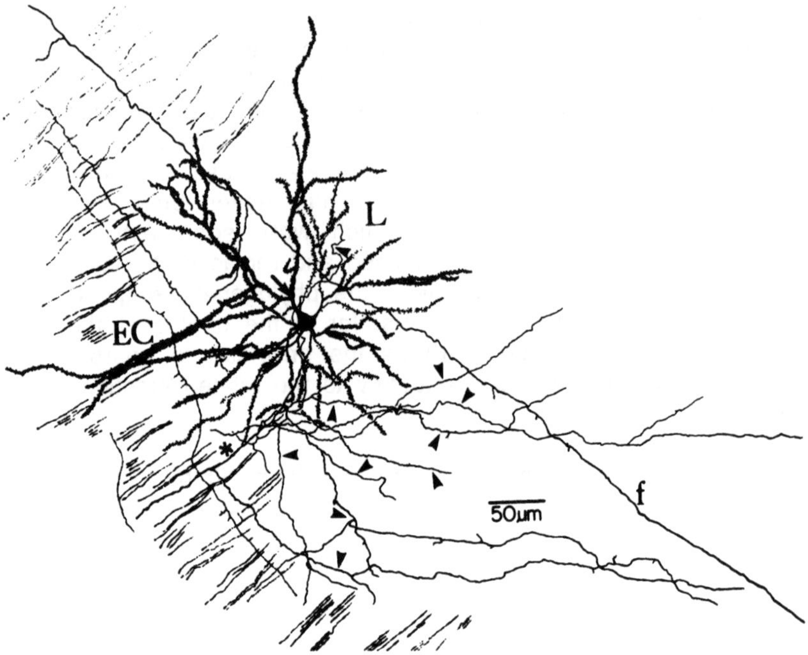

Fig 1. Typical pyramidal cell in Golgi preparation from the far lateral amygdala (L) of a 3-month-old (adult) mouse. Note that the dendrites are covered with a profusion of spines. The cell body is about 18 µm in diameter, and the major dendrites >250 µm in length. The apical dendrite is somewhat unusual in being bifurcated (arrow), one branch of which extends into the external capsule (EC). The principal axon (asterisk) also passes into EC; and numerous axon collaterals (arrow heads) project dorsally, ventrally and medially. A second axon, f, from some other source traverses this segment of the lateral amygdala, giving off boutons en passant. (From Millhouse and DeOlmos 1983, with permission)

AMYGDALA

As will be outlined below, a very substantial body of evidence points to a central role for the amygdala in the expression, and probably the experience, of emotion in man and animals. How it is that this assemblage of neurons is able to generate such intense and ineffable subjective states is perhaps no more mysterious than how the retinocortical system generates the miracles of sight; yet it remains a profound challenge to discern whether the quality of experience is dependent upon the idiosyncracies of the neuronal action within a particular neural entity, or whether it is, fundamentally, the connectivity of the system which yields the experiential outcome. In so far as the amygdala is concerned, it would appear that its uniqueness does not lie in the nature of its neuronal elements, for at least in the Golgi picture the neurons which comprise it bear striking similarity to those of the cerebral cortex (fig. 1), e.g., pyramidal cells with highly spinous "apical" and basilar dendrites, stellate cells, and chandelier cells(McDonald 1982; Millhouse and DeOlmos 1983); or in the central nucleus, to cells of the putamen or caudate nucleus (Hall 1972).

Despite this rather uniform nature of the basic elements, the various nuclear groupings within the amydgala are extraordinarily complex in their afferent and efferent connections, and in the distribution of various chemical moieties within them, such as acetylcholinesterase, zinc, the monoamines, and peptides. In the amygdala of rodents the sex steroids produce definitive histological and histochemical patterns (Raisman and Field 1972; Arimatsu et al. 1981; Nishizuka and Arai 1981); but such effects have not yet been analyzed in primates. It is also worthy of note that although the Golgi patterns suggest only the presence of neurons comparable to those in neocortex (fig. 1), the amygdala is uniquely prone to the development of seizure activity, differing markedly from neocortex in this regard (Pagni and Maroserro 1965; Goddard 1972; Overman and Doty 1979; Wada et al.1981).

Afferent and efferent fibers of the amygdala follow three mayor patways in primates (Nauta 1961): (1) a major pathway which arises rostrally as a continuation of the intra amygdaloid, longitudinal association bundle of Johnston and which courses ventrally into the preoptic area, its mass approximating that of the anterior commissure although it is much more dispersed. (2) A diffuse pathway which connects the amygdala laterally and ventrally with neo and allocortex. (3) The stria terminalis which takes origin primarily from the caudal components and which is considerably reduced in primates as compared to other mammals. The latter is accompanied in its looping course beneath the caudate nucleus by clumps of cells, the bed nucleus of the stria terminalis, which are reminiscent of the islets of Calleja, and which ultimately merge with the bed nucleus of the anterior commissure as the stria disperses around this structure into the preoptic area and hypothalamus.

Present knowledge of the afferent and efferent connections of the amygdala has recently been summarized by Amaral (1987), Russchen (1986), and Price (1986), from whom the following outline has been abstracted, with particular reference to primates. The latter qualification is important, since the amygdala of macaques is endowed with a rich interconnectivity with neocortex (fig. 2) that is, at best, poorly developed in non primate species.as can be discerned in fig. 2.

The neocortical afferents arise from the insula and orbitofrontal cortex, and particularly from the visual areas of the temporal lobe (Iwai and Yukie 1987). In addition, the entorhinal cortex projects upon the amygdala. The major sources of subcortical afferents are provided by elements of the limbic system, especially from the hypothalamus, e.g., the ventromedial and arcuate nuclei. The four modulatory systems, i.e., cholinergic from nucleus basalis and the diagonal band, dopaminergic from the ventral tegmental area and substantia nigra pars compacta, noradrenergic from locus coeruleus, and

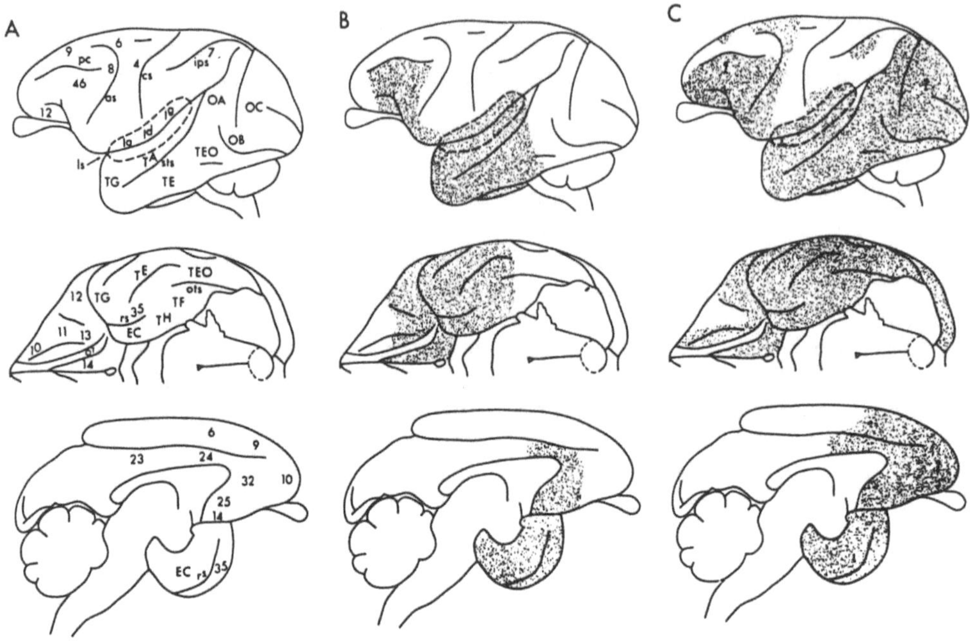

Fig 2. Cortical areas (A) in macaques that send (B) or receive (C) axons directly to or from the amygdala, as shown by shaded areas in B and C. Note the very extensive interchange between amygdala and neocortex. (From Amaral 1987, with permission)

serotonergic from the dorsal raphe and nucleus centralis superior project conspicuously into the amygdala, as do the olfactory and gustatory systems (the latter via peribrachial nucleus). Finally, there is a substantial input from the medial pulvinar and a scattering from midline thalamic nuclei the neural groups within the amygdala reciprocate these afferent connections to various degrees. Perhaps most remarkable in macaques is the projection into the temporal lobe, attaining primary auditory cortex (Fig. 2), and continuing back into the circumstriate and even striate cortex (Mizuno et al. 1981; Tigges and Tigges 1985; Amaral 1987; Iwai and Yukie 1987 - Fig. 2). Macaques are also differentiated from other species by the strong projection of the amygdala into many of the components of the hippocampal system (Amaral 1987; Rosene and Van Hoesen 1987). The amygdala provides a major source of afferents to nucleus basalis (Aggleton et al. 1987); and projects heavily onto the preoptic area and hypothalamus, and onto the mediodorsal nucleus of the thalamus (which, in turn, is a major afferent source for the frontal lobe). Other efferents pass to the striatum, the midbrain, and even down to the dorsal nucleus of the vagus and nucleus of the solitary tract.

The amygdala shares with the hippocampus in primates (Rosene and Van Hoesen 1987) a sparsity or absence of interhemispheric connections. While there is occasional evidence of a meager projection from one amygdala to the other across the anterior commissure (Iwai and Yukie 1987), others have failed to find it (Fox et al. 1949; Nauta 1961; Pandya and Seltzer 1986; Demeter et al, in preparation). There is perhaps some individual variability here, since the human anterior commissure, at least, displays a sevenfold range in cross sectional area (Demeter et al. 1988). Yet the degeneration of fiber terminals in the amygdala after transection of the anterior commissure in macaques (Turner et al. 1979) may well reflect input from the contralateral neocortex, or even subcortical afferents. In any event, the electrical record in one amygdala is independent of that in its contralateral mate in both man and macaque, even in regard to propagation of seizure activity (Poblete et al. 1959; Pagni and Marossero 1965; Cabrini et al. 1970; Brazier 1972; Wada et al. 1981). Furthermore, if the ventral amygdalofugal path, 1) above, is interrupted ipsilateral to stimulating electrodes in the amygdala, the latter are rendered incapable of eliciting the defense reaction which they previously evoked (Fernandez de Molina and Hunsperger 1962; Hilton and Zbrozyna 1963); i.e., an effective path into the contralateral amygdala is lacking.

The unilaterality of this anatomical situation has important functional implications, for if the amygdala is a primary source or requirement for certain emotions, then such emotion is initially private to a given hemisphere and is shared with the other only through secondary, probably neocortical pathways. That such sharing is possible will be discussed anon; but should such "sharing" be impaired, the anatomical possibility also exists for each hemisphere to pursue a semi-independent emotional life!

That some mechanism in the temporal lobes is uniquely related to emotional expression has been clear since the observations of Brown and Schäfer, and Klüver and Bucy, and the fact that emotional or even motivational (e.g., Doty and Giurgea 1961; Doty 1969) effects are lacking at other cortical loci. The exact nature and location of that emotional substrate still remains unclear, but a substantial body of evidence points to the amygdala as the most likely and primary focus of the temporal lobe effects. Although modern comparative anatomists, e.g., Northcutt (1981) and Parent (1986), profess a certain skepticism in homologizing between the mammalian amygdala and somewhat similarly positioned cell groups in nonmammalian species, destruction or electrical stimulation of the putative amygdala but not adjacent regions in reptiles (Keating et al. 1970; Tarr 1977) produce dramatic reductions or increases, respectively, in aggressive behavior. It is thus likely that the amygdala has a long phylogenetic history as a locus for the evaluation of environmental stimuli. This is dramatically evidenced by the disappearance of the unrestrained viciousness in such persistently intractable animals as the wild rat (Woods 1956) or lynx (Schreiner and Kling 1956) following amygdalectomy; just as with the tameness similarly observed in Brown and Schäfer's macaques and the Klüver-Bucy syndrome. That this unusual docility is attributable to loss of the amygdala in the latter cases is now well documented (Clemente and Chase 1973), and has in fact provided the rationale for amygdalectomy in certain destructive and aggressive human patients (see Valenstein 1973). Equally convincing are the effects of electrical stimulation of the amygdala, which commonly unleashes a fully developed attack or defense response in animals (e.g., Fernandez de Molina and Hunsperger 1962; Hilton and Zbrozyna 1963; Clemente and Chase 1973).

Probably because of the powerful effects of social control, attack behavior is rarely elicited by electrical stimulation of the amygdala in human subjects; but fear is a common accompaniment (e.g., Halgren et al. 1978; Gloor et al. 1982). However, for a surprising number of loci, stimulation may produce no subjective effect (Pagni and Maroserro 1965; Halgren et al.1978; Gloor et al. 1982), and those which are produced are extremely diverse, and tend to be idiosyncratic. This is, first of all, undoubtedly attributable to the great variety of inputs and the complexity of the amygdaloid complex itself; but more subtle effects are probably also at work. These would reflect the stamp of individual experience and inheritance, that can profoundly influence the effects obtained by electrical activation of particular neural circuits, even as in the "personality" of Rioch and Brenner's cats (1937), or the deliberately altered electrically elicited responses of Valenstein et al. (1970) or Doty and Giurgea (1961).

Two important features of the amygdala in relation to emotion receive emphasis in a series of instructive observations by Fonberg (1972): the fact that the amygdala is actually a complex of nuclei subserving distinguishably different functions, and the powerful control which the amygdala exerts over the hypothalamus. She showed that lesions in the lateral amygdala would restore the vivacious behavior of dogs rendered apathetic and aphagic by lesions in the lateral hypothalamus. In similar vein, following lesions in the dorsomedial amygdala, dogs were deeply "depressed" and aphagic, comparable to those with lesions of the lateral hypothalamus; and in this case also, extension of the destruction to include the remainder of the amygdala, i.e., to the lateral component, restored them to a condition approaching more normal behavior. This line of experimentation was subsequently extended to show that loss of the dorsomedial amygdala or lateral hypothalamus strongly perturbed a dog's well developed social bond to human individuals (Fonberg and Kostarczyk 1980).

The link between social perceptiveness and the amygdala is strikingly apparent in primates. Underlying this is the predominance of the visual modality in primate social interaction, and the strong projection of higher levels of visual cortical analysis in the inferior temporal lobe into the amygdala (e.g., Iwai and Yukie 1987). The activity of occasional single neurons in the amygdala of alert macaques can reflect this input in a remarkably specific manner, such as responding only to particular faces of monkey or man, or even preferentially to the faces of other monkeys displaying a threat (Leonard et al. 1985). Other units may respond in a highly selective manner only to a particular threatening object, such as a hypodermic syringe, or parts thereof; and such response may be inhibited if the syringe is shown to the monkey together with food, but not if with inedible objects (Ono et al. 1983). It bears further note, however, that such meaningful visual effects in the amygdala may not be limited to primates, for single unit activity in the deep temporal cortex of sheep have now also been shown which is highly selective for specific visual inputs, such as faces or heads of other sheep, human beings or dogs (Kendrick and Baldwin 1987).

Interruption of the path conveying input from the visual cortical areas into the amygdala can thus, understandably, produce some aspects of the Klüver-Bucy syndrome (Horel and Keating 1975). The occurrence of this syndrome in man (e.g., Lilly et al. 1983), however, is still further complicated by the vagaries of the pathology; nor is the Klüver-Bucy syndrome necessarily the outcome of such elision between visual cortex and amygdala. Bilateral disconnection of the amygdala from temporal neocortex in a male patient has produced a singular apathy to visually perceived sexual stimuli without detriment to arousal by sexually explicit verbal descriptions; along with a severe impairment of the visual appreciation of faces, such as the subject's enduring inability to recognize his father visually (Bauer 1982).
Thus, the amygdala in primates, to usurp a phrase from Sherrington, serves as the head ganglion of the emotional-motivational system, receiving and returning substantial connections with neocortex to extract sensorial information and, in similar exchange with the hippocampus, evoking and/or recording mnemonic data, and finally, taking account of these sources, commandeering the hypothalamus and the autonomic system for suitable motor or hormonal output.

HABENULO-INTERPEDUNCULAR COMPLEX

The anatomy of this diverse assemblage of nuclei and tracts is reasonably well delineated, as perceptively outlined by Jones (1985). They form a major component of the subcortical limbic system. Fibers gather from the septal, diagonal band and preoptic areas, and the globus pallidus, to form the stria medullaris (Fig. 3). These fibers distribute to the habenular nuclei, but some continue on to the interpeduncular nucleus and tegmentum. A number of them decussate in the habenular commissure. Cells of the medial habenular nucleus, small and closely packed, receive input only from the septal region, and project via the fasciculus retroflexus (habenulopeduncular tract) to the interpeduncular nucleus, where their fibers make an extraordinary "figure 8", looping from one side to the other and back again. Cells of the lateral nucleus of the habenula are larger and more dispersed, and receive input from the preoptic area, lateral hypothalamus, diagonal band, pallidum, and nucleus basalis. They too project into the stria and down the fasciculus retroflexus, but exit prior to the interpeduncular nucleus to provide the major source of afferents to the dorsal and median raphe (Fig. 4) which, in turn, constitute the ascending serotonergic system. The lateral nucleus also supplies fibers to the substantia nigra and ventral tegmental area, and returns a diffuse projection into the preoptic and septal region (see Jones 1985).

The interpeduncular nucleus is composed of a number of subnuclei, and is distinguished by having an exceptionally rich and varied concentration of peptidergic and other receptors (e.g., Groenewegen et al. 1986). The cholinergic enzymes are notably prominent. They are heavily but not

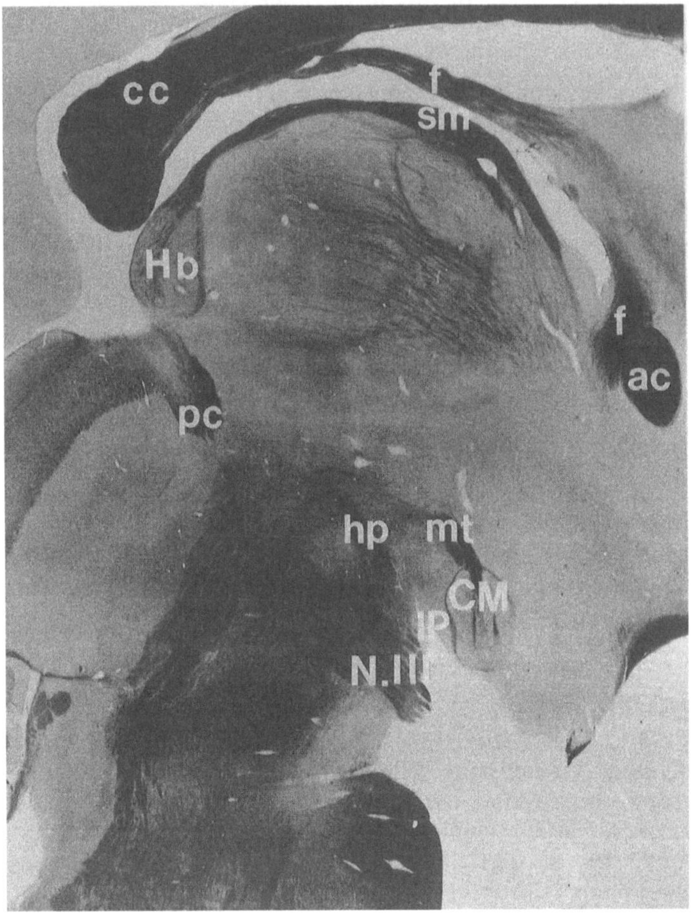

Fig 3. Fiber stained, sagittal section 1 mm lateral to the midline in a juvenile (2.5 Kg) male Macaca nemestrina, showing elements of the habenulo-interpeduncular system. The fasciculus retroflexus (habenulopeduncular tract, hp) can be seen as it cuts through the red nucleus. The habenular complex (Hb) is divided into medial and lateral nuclei (rostral and caudal, respectively, in this perspective) by fibers of the stria medullaris (sm). The interpeduncular nucleus (ip) is surrounded by rootlets of the oculomotor nerve (N.III). Other labels as in Fig. 4, except pc, posterior commissure.

exclusively dependent upon input from the habenula (e.g., Clarke et al. 1986; Eckenrode et al. 1987). The pattern of projection upon the various subnuclei is complex, some receiving bilateral, symmetrical input from the fasciculus retroflexus, others having predominantly ipsilateral or contralateral inputs, a large percentage of the fibers being cholinergic (from the ventral portion of the medial habenular nucleus), whereas some subnuclei receive fibers labeled for substance P (Groenewegen et al. 1986; Eckenrode et al. 1987).

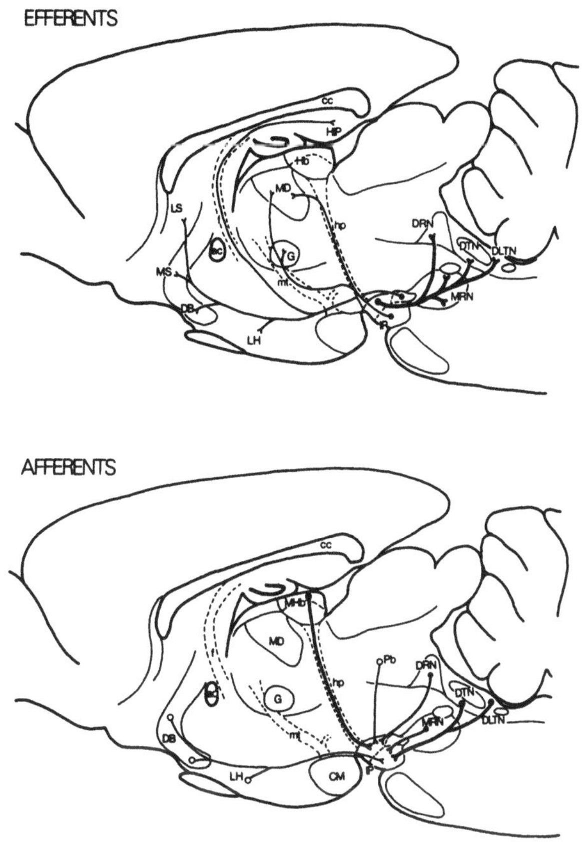

Fig 4. Diagram of the efferent and afferent connections of the interpeduncular nucleus in the rat. Code: ac, anterior commissure; cc, corpus callosum; CM, corpus mammillare; DB, nucleus of the diagonal band of Broca; DTLN, nucleus tegmenti dorsalis lateralis; DRN, nucleus raphè dorsalis; DTN, nucleus dorsalis tegmenti of Gudden; f, fornix; G, nucleus gelatinosus thalami; Hb, habenula; HIP, hippocampus; hp, tractus habenulo-interpeduncularis; IP, nucleus interpeduncularis; LH, lateral hypothalamus; LS, nucleus septalis lateralis; MD, nucleus mediodorsalis thalami; MHb, medial habenula; MRN, nucleus raphè medialis; MS, nucleus septalis medialis; mt, tractus mammillothalamicus; Pb, parabigeminal nucleus. (From Groenewegen et al. 1986, with permission)

The interpeduncular subnuclei, like the fasciculus retroflexus, send a strong projection, with intricate sorting of laterality, into the raphe and into the tegmental nuclei. A small contigent also passes to the central gray, and to the mediodorsal nucleus of the thalamus and septal area, returning in large measure via the fasciculus and the stria medullaris. The prominent efferent projection into raphe and tegmentum is reciprocated by afferents from these nuclei, again involving an intricate sorting of ipsi-and contralateral paths into the various interpeduncular subnuclei (Groenewegen et al. 1986 - Fig. 4).

It is apparent from the foregoing that the habenulo-interpeduncular complex is intimately involved with control of the serotonergic system (Azmitia and Gannon 1986) and other fibers ascending from the raphe and tegmental nuclei. However, the direction and nature of this control is still a matter of controversy (see Motohashi et al. 1986).

Even more challenging, of course, is the fact that the functional significance of the entire system remains wholly mysterious (see, e.g., Sutherland 1982). Perhaps this were reason enough to include this system in the present review, to draw attention to a probably important realm of ignorance. Yet there are other features which add to the relevance of such discussion. First is the unexpectedly bilateral nature of the projection of the raphe serotonergic system in macaques with the possibility, so far unexamined, that this projection is asymmetrical between the hemispheres. Following unilateral injections of horseradish peroxidase into striate cortex, 25%-45% of the labeled neurons in the raphe system were contralateral (Doty 1983), and similar data were obtained for injections into the hippocampus (Amaral and Cowan 1980). Insofar as it has been examined, the contralateral component in the rat and cat seems much less than in primates. The meaning of such bilateral organization, particularly if under control of the intricate laterality seen in the projections to and from the interpeduncular subnuclei, is something of a Gordian knot, but one likely bearing some relation to the asymmetry of hemispheric effects and affect. Mandell and Knapp (1979) reported a roughly 20% asymmetry in serotonin-containing neurons on one versus the other side of the raphe in rats, along with an asymmetry in serotonin concentrations in striatum and hippocampus; but there was apparently no consistency as to which side was high.

Along these lines it is of interest to note that the habenula displays the most spectacular right-left asymmetry of any structure in the vertebrate brain (see Braitenberg and Kemali 1970 for brief review); and although this asymmetry has been thought to be limited to fish, its recent rediscovery in amphibians and now birds (Gurusinghe et al. 1986) suggests that its phylgenetic history may not end so soon. The latter observation is particularly intriguing, since the asymmetry is sex dependent and under the control of testosterone. The habenulae are of equal size in normal female chickens, but in males, or females treated with testosterone, the right is some 6%-10% larger than the left; the direction, and the equality in females, both being concordant with what might be expected for a gross extrapolation from numerous studies of hemispheric asymmetry in man (McGlone 1980).

In the frog there is clear evidence for an olfactory influence on the medial habenular nuclei since they are somewhat atrophied by removal of the olfactory pits during early development; but the unpaired lateral habenular nucleus was unaffected (Norris and Eichler 1975). Destruction of the habenulae in the rat also produces a distinct impairment in its ability to utilize olfactory information for choosing a path to food, while its visual choice is probably unaffected (Rausch and Long 1971).

A more consistently pursued thread is that the habenular system has something to do with temperature regulation. To begin with, major input to the lateral habenular nucleus in mammals arises in the preoptic area and globus pallidus (Jones 1985; Parent 1986). The former is a well recognized center for the regulation of body temperature; and stimulation throughout a reasonable approximation of these areas by Schaltenbrand et al. (1973) rather commonly evoked in their patients a sensation of general warmth, often accompanied not only by objective evidence of cutaneous vasodilation but by a sense of euphoria and well being! Stimulation ventral to the anterior commissure, on the other hand, was more likely to evoke a feeling of being cold. In an extensive study using anesthetized rabbits Cragg (1961) was able to elicit panting by local heating or electrical excitation of the preoptic area. Heating was ineffective at other loci, but electrical stimulation applied throughout the system could elicit panting, from the habenula, interpeduncular, and dorsal tegmental nuclei. When the stria medullaris was transected, unanesthetized rabbits no longer displayed a hyperpnea upon restraint, thus possibly indicating emotional as well as "temperature" control via the habenular system.

In addition to such phasic regulation of body temperature, there is a somewhat tentative relation with thyroid function. Thyrotropin releasing hormone is present in the preoptic area, and electrical stimulation there can induce the release of thyroid stimulating hormone (Martin and Reichlin 1987). A variety of observations further demonstrate an interaction between the habenula and thyroid function (Szentàgothai et al. 1962). This is especially clear in the quail, where loss of the habenulae decreases plasma thyroxine almost 50% (Herbuté et al. 1977). Such relations with thyroid function are perhaps sequelae of a phylogenesis in which the habenulae were intimately related with the pineal body, a relation which is still demonstrable. Electrical pulses applied to the habenula can drive 17% of the single units encountered in the pineal body of the anesthetized rat (Ronnekleiv et al. 1980); and it is possible that some of the stimulated fibers take origin from the preoptic area (Moller and Korf 1983).

Effects in the behaving animal are even more difficult to interpret. In the few cases assayed, direct electrical stimulation of the habenula or interpeduncular nucleus in cats (Nielson et al. 1958; Doty 1961) was highly aversive; and the same seemed to be true of the habenula in the rat (Olds and Olds 1963). Reinert (1964) also stimulated the habenula in alert cats and noted "a pattern one could associate with a frightened animal, cowering and withdrawing slowly," in addition to vasodilation in muscle for cats anesthetized with chloralose. Perhaps as a further example of species differences in relation to electrical self-stimulation of the brain (Doty 1969), Sutherland and Nakajima (1981), while observing some initial hesitancy, found that rats would ultimately achieve a modest level of self-stimulation of the habenula or fasciculus retroflexus. The effect seemed to be dependent upon intactness of the median raphe. Rats with lesions of the habenula (Nielson and McIver l966; Wilson et al. 1972), interpeduncular nucleus (Wilson et al. 1972), or transection of the connections between the interpeduncular nucleus and the tegmentum (Wirtshafter 1981) behave similarly to animals with lesions of the septal area in that they have trouble suppressing an inappropriate response, although they do not display a pronounced hyperactivity. They are, however, again like animals with septal lesions, more active and therefore more successful in learning to avoid an impending shock by moving to another location. Yet such animals are deficient in selecting more complex behavioral strategies for escape (Thornton and Evans 1982).

Thus, the emotion of "warmth" or "well-being" (Schaltenbrand et al. 1973) is perhaps as close as one can come to reasonable speculation about the emotional concomitants of the habenulo-interpeduncular system and that, obviously, on a very tenuous basis. However, the close but complex link with the ascending serotonergic system, and even with the central gray, suggests that many facets of emotional life may lie still undeciphered within this phylogenetically long-enduring network. Indeed, in their review Mandell and Knapp (1979) cite various pieces of evidence for participation of the serotonergic system in the modulation of "attention, sleep, temperature, pain, aggression, activity level, self-stimulation, sexual behavior, avoidance learning, water consumption, and appetite"!

EMOTION AND THE TWO HEMISPHERES

As plentifully presented in this volume and elsewhere (e.g., Gainotti 1987; Otto et al. 1987), there is substantial evidence that the two cerebral hemispheres in man often differ in their emotional propensities. A dramatic example of this asymmetry is seen in panic disorder, in which the right hemisphere shows significantly greater effects than the left (Reiman et al. 1986; Buchsbaum et al. 1987).

There are no obvious clues as to why such an arrangement should prevail. However, a number of potentially fruitful speculations can be offered. First is the fact noted above that the amygdala, as the principal organizer of emotional reactions, is decidedly unilateral in its anatomical relations and electrophysiological manifestations. Thus, each hemisphere is initially isolated from the emotional appraisals in the other, and intercommunication at the emotional level must proceed via neocortical interchange across the forebrain commissures. There are, of course, the convolved crossings and recrossings of the habenulo-interpeduncular complex and its serotonergic-raphe companion, which might convey subtle emotional shadings from one hemisphere to the other (see Sperry et al. 1979; Cronin-Golomb 1986) but not an accurate depiction of their source. In any event, the neocortex can control or override subcortical emotional effects, as can be clearly seen in the experiments of Cytawa and Teitelbaum (1968). In rats that had fully recovered from the bizarre hyperemotionality that follows lesions of the septal area, cortical spreading depression reprecipitated the syndrome and the lengthy course of recovery from it (2-3 weeks).

One must thus look ultimately to the cerebral cortex for the origin of hemispheric disparity, and it seems likely that the explanation lies in the fact that the two hemispheres are continually in a state of dynamic equilibrium and competition, both anatomically and functionally. Pearson et al. (1985) found that transection of the corpus callosum even in adult rats produced a bilateral increase in size of cholinergic neurons in nucleus basalis (whereas the size of these neurons diminishes if their axons are damaged). Presumably, the increase in size of these neurons reflects their success in competing for the considerable synaptic space in each hemisphere vacated by the loss of callosal input. A substantial body of data, reviewed by Innocenti (1986), shows that the interconnections of the corpus callosum are those which prevail in the face of such anatomical competition for synaptic space, which in turn is under the control of environmental, sensorial input. The callosal paths in man are not fully myelinated until 12 years of age (Yakovlev and Lecours 1967) and, indeed, there is evidence of a partial "split-brain" syndrome in 4-year old children (Joseph et al. 1984). One can readily imagine, then, that as the synapses in the left hemisphere are pre-empted in the service of language, those in the right develop different patterns. The relative independence of development of the two hemispheres can be detected electrophysiologically (Thatcher et al. 1987).

The impetus towards hemispheric specialization is likely to arise from the extraordinarily long integration times for multisynaptic processing were this to be achieved in a brain the size of that in man via interhemispheric connections (JL Ringo and RW Doty, in preparation). The basic facts are set forth by Swadlow et al. (1979), interhemispheric conduction times being on the order of l00-200 ms for most callosal axons in man; and for the slowest primate axons following some of the longer possible routes in the human brain the time for a one-way trip is almost one-third of a second!

Interhemispheric competition is manifested in a variety of ways, as reviewed by Cook (1986). One of the more interesting is the unilaterality of learning when the cue is initiated by electrical excitation of striate cortex in macaques (Doty et al 1973); i.e., despite intactness of callosal connections during training, only one hemisphere learns the response, thus suggesting that the other is held in check by that hemisphere from which the critical information initially emanates. A more specific balance between the hemispheres, particularly relevant to emotion, is seen in the experiments of Colle and Wise (1987). They showed that extirpation of rostral forebrain in rats significantly raised the threshold for electrical self-stimulation of the median forebrain bundle ipsilateral to the lesion, and decreased it for contralateral stimulation. The contralateral decrease was permanent, while the ipsilateral threshold returned to its preoperative level within about 14 weeks.

Although it has been emphasized above that the amygdala lacks direct input to the contralateral hemisphere, it has also been shown that its emotional contribution to the contralateral hemisphere can be effected via the corpus callosum (or, specifically, the splenium of the corpus callosum). Downer (1961) developed the means of demonstrating this. In a macaque with an amygdalectomy on one side and the optic chiasm transected, the animal's behavior is normal when it views the world with either eye, so long as the corpus callosum is intact. But when transection of the corpus callosum (and anterior commissure, cut during the amygdalectomy) is added, the peculiar situation arises in which the animal is tame and insouciant when viewing the world through the eye on the side of the amygdalectomy; yet displays its usual wild, frightened and aggressive behavior when its view of the world is through the eye serving the hemisphere still having an amygdala. Thus, the "same monkey" has a completely different emotional outlook depending upon which eye and hemisphere it uses to appraise its environment. Downer's observations have been confirmed by Barrett (1969), Horel and Keating (1972), and Doty et al. (1973).

Our refinements were to cut the optic tract in one hemisphere, rendering it "blind" but retaining its amygdala, and to perform the amygdalectomy in the other hemisphere. All of the callosum was then cut except the splenium, which was captured in a snare that could be used to complete the transection under local anesthesia. A macaque so prepared is wholly normal in its fear of man until the instant the splenium is transected, after which it displays the characteristic lack of fear to all objects formerly threatening. Thus, the amygdala in a "blind" hemisphere can supply the emotional evaluation to visual input garnered via the other; the question is, what is the direction of this informational exchange, i.e., does the seeing hemisphere send the amygdala some abstract of the visual scene, or does the amygdala provide the visual system continually with a series of templates or filters for relevant targets? Since cells of the amygdala "recognize" highly specific patterns at a surprisingly short latency (Leonard et al. 1985), it might be supposed that the former is the more likely; yet the amygdala itself sends projections into the visual cortex (Mizuno et al. 1981; Tigges and Tigges 1985; Amaral 1987; Iwai and Yukie 1987), and could thus serve in the second capacity. These experiments demonstrate, albeit in a somewhat peculiar sense, that even though the amygdala may operate basically as a unilateral system, its emotional effects can be communicated between the hemispheres via the corpus callosum. It must thus be asked how, in human beings, one hemisphere may have a different emotional outlook than the other? The answer would seem to lie with the hemispheric specialization which proceeds despite (indeed, probably because of - see Doty et al. 1973) the presence of the interconnecting callosal system. And there is, of course, the added possibility that in the event the callosal mechanism is awry, the emotional evaluations of a particular hemisphere may exist relatively independently of the emotional state in the other, or come to dominate it.

This is not to say that one hemisphere has no subcortical access to emotional reactions generated within the other hemisphere. Sperry et al. (1979) demonstrated in patients with surgically transected forebrain commissures that the emotional concomitants evoked in the right hemisphere by visual presentation of personally relevant material were communicated to the nonseeing left hemisphere, even though the latter remained nescient as to the exact nature of the material. Cronin-Golomb (1986) has gone on to show that such interhemispheric communication of emotional tone can have a remarkable precision, while nevertheless remaining wholly nonexplicit in detail. This emphasizes the subtlety and almost infinite shading of emotional tone, which may be a frequent accompaniement of otherwise purely cognitive processes.

On the other hand, intellectual decisions, patently lacking in differential emotional coloring, can be achieved cooperatively via subcortical mechanisms, integrating information supplied to each hemisphere individualy, yet again unavailable to either in specific detail (Sergent 1986).

Gott et al. (1984) reported a remarkable case of a woman who as an adolescent had been disturbed by involuntary shifts in emotional and intellectual tone, and who had subsequently been able to bring these alterations of state under voluntary control. Analysis of EEG and a variety of tests for hemispheric specialization convincingly showed that in these voluntary alterations of her subjective state the subject was actually shifting dominance from one to the other hemisphere. This was accompanied subjectively by a distinct change in mood or emotional tone. How such shifts might be achieved, whether through the forebrain commissures or via cortical control of the balance among subcortical systems, or both, is a matter of prime significance in understanding how two potentially independent entities, the two cerebral hemispheres, unite in the production of singularly undivided conscious and emotional being; or how they might interact to produce mental illness (Doty 1988).

IN SUMMARY

1. It is emphasized that elementary emotional expression and "personality" survive total loss of neocortex and that, therefore, emotion is essentially a product of subcortical processes.

2. The amygdala is the "head ganglion" of this subcortical emotional system. It is in possession of complex sensorial analysis in primates, as supplied by substantial input from neocortex, and returns extensive connections thereto; is richly provided with the capacity to exchange information with the hippocampus, probably in reference to recall or recording of past events; and, ultimately, controls the hormonal and autonomic apparatus of the hypothalamus and the motor systems of the midbrain to effect appropriate output.

3. Both anatomical and electrophysiological evidence indicate that the amygdala is uniquely unilateral in its operation and that, therefore, its contribution to emotion is initially private to a given hemisphere.

4. The neocortical constituents of the hemispheres can, however, exchange emotionally relevant information via the forebrain commissures. A surprisingly subtle, if nonspecific, interhemispheric interchange of

emotionally toned information can also proceed over exclusively subcortical channels.

5. Via the forebrain commissures the two hemispheres are in a constant state of functional and anatomic competition, a process probably underlying hemispheric specialization both in the intellectual and in the emotional realm.

6. The enigma of the habenulo-interpeduncular complex is reviewed, pointing out: (a) its control of and interrelationship with the ascending serotonergic system of the raphe, and (b) its phylogenetic background of extreme lateral asymmetry.

Acknowledgements Preparation of this review was supported by a Javits Award, NS20052, from the National Institute of Neurological and Communicative Disorders and Stroke, National Institutes of Health. I am grateful to my colleagues Steven Demeter, James L.Ringo, and Jeffrey D. Lewine for a continual flow of ingenious and provocative ideas in our discussions of topics dealt with herein.

REFERENCES

Adamec RE, Stark-Adamec C, Livingston KE (1980) The development of predatory aggression and defense in the domestic cat (Felix catus). Behav Neural Biol 30: 389-409

Adams DB (1979) Brain mechanisms for offense, defense and submission. Behav Brain Sci 2:201-241

Aggleton JP, Friedman DP, Mishkin M (1987) A comparison between the connections of the amygdala and hippocampus with the basal forebrain in the macaque. Exp Brain Res 67: 556-568

Amaral DG (1987) Memory: anatomical organization of candidate brain regions. In: Mountcastle VB, Plum F, Geiger SR (eds) Handbook of physiology, Sect 1: The nervous system, Vol V. Higher functions of the brain, Part 1, American Physiological Soc, Bethesda, Maryland, 211-294

Amaral DG, Cowan WM (1980) Subcortical afferents to the hippocampal formation in the monkey. J Comp Neurol 189: 573-591

Arimatsu Y, Seto A, Amano T (1981) Sexual dimorphism in alpha-bungarotoxin binding capacity in the mouse amygdala. Brain Res 213: 432-437

Azmitia EC, Gannon PJ (1986) The primate serotonergic system: a review of human and animal studies and a report on Macaca fascicularis. In: Fahn S, Marsden CD, van Noort MH (eds) Advances in neurology, Vol 43: Myoclonus, Raven, New York, pp 407-468

Baleydier C, Mauguiere F (1980) The duality of the cingulate gyrus in the monkey; neuroanatomical study and functional hypothesis. Brain 103: 525-554

Bandler R, Prineas S, McCulloch T (1985) Further localization of midbrain neurones mediating the defence reaction in the cat by microinjections of excitatory amino acids. Neurosci Letters 56:311-316

Bard P, Rioch D McK (1937) A study of four cats deprived of neocortex and additional portions of the forebrain. Bull John Hopkins Hospital 60: 73-147

Barrett TW (1969) Studies of the function of the amygdaloid complex in Macaca mulatta. Neuropsychologia 7:1-12

Bauer RM (1982) Visual hypoemotionality as a symptom of visual-limbic disconnection in man. Arch Neurol 39:702-708

Bertrand M (1969) The behavioral repertoire of the stumptail macaque. Karger, Basel,

Bignall KE, Schramm L (1974) Behavior of chronically decerebrated kittens. Exp Neurol 42:519-531

Braitenberg V, Kemali M (1970) Exceptions to bilateral symmetry in the epithalamus of lower vertebrates. J Comp Neurol 138:137-146

Brazier MAB (1972) The human amygdala: electrophysiological studies. In: Eleftheriou BE (ed) The neurobiology of the amygdala. Plenum, New York, pp 397-420

Brown S, Schäfer EA (1888) An investigation into the functions of the occipital and temporal lobes of the monkey's brain. Phil Trans Royal Soc 179B: 303-327

Buchsbaum MS, Wu J, Haier R, Hazlett E, Ball R, Katz M, Sokolski K, Lagunas-Solar M, Langer D (1987) Positron emission tomographe assessment of effects of benzodiazepines on regional glucose metabolic rate in patients with anxiety disorder. Life Sci 40: 2393-2400

Bucy PC, Klüver H (1955) An anatomical investigation of the temporal lobe in the monkey (Macaca mulatta). J Comp Neurol 103:151-251

Cabrini G, Ettorre G, Infuso L, Marossero F (1970) Connessioni delle strutture rinencefaliche nell'uomo. Osservazioni raccolte in corso di esplorazioni SEEG. Riv Neurol 40:249-255

Carpenter CR (1964) Naturalistic behavior of nonhuman primates. Pennsylvania State Univ Press, University Park,

Clarke PBS, Hamill GS, Nadi NS, Jacobowitz DM, Pert A (1986) alpha-nicotine-and-alpha-bungarotoxin-labeled nicotinic receptors in the interpeduncular nucleus of rats. II. Effects of habenular deafferentation. J Comp Neurol 251:407-413

Clemente CD, Chase MH (1973) Neurological substrates of aggressive behavior. Ann Rev Physiol 35: 329-356

Colle LM, Wise RA (1987) Opposite effects of unilateral forebrain ablations on ipsilateral and contralateral hypothalamic self-stimulation. Brain Res 407:285-293

Cook ND (1986) The brain code. Mechanisms of information transfer and the role of the corpus callosum. Methuen, London

Cragg BG (1961) The role of the habenula in the respiratory response of the rabbit to warmth or to restraint. Exp Neurol 4:115-133

Cronin-Golomb A (1986) Subcortical transfer of cognitive information in subjects with complete forebrain commissurotomy. Cortex 22: 499-519

Crusio WE, Schwegler H (1987) Hippocampal mossy fiber distribution covaries with open-field habituation in the mouse. Behav Brain Res 26:153-158

Cytawa J, Teitelbaum P (1968) Spreading depression and recovery from septal hyperemotionality. Folia Biologica 16:459-468

Demeter S, Ringo JL, Doty RW (1988) Morphometric analysis of the human corpus callosum and anterior commissure. Human Neurobiol 6:219-226

Demeter S, Rosene DL, Van Hoesen GW (In preparation) The fields of origin and pathways of the interhemispheric commissures in the temporal lobe of macaques

Doty RW (1961) Conditioned reflexes formed and evoked by brain stimulation. In: Sheer DE (ed) Electrical stimulation of the brain. Univ Texas Press, Austin, 397-412

Doty RW (1969) Electrical stimulation of the brain in behavioral context. Ann Rev Psychol 20:289-320

Doty RW (1976) The concept of neural "centers". In: Fentress J (ed) Simple networks. Sinauer, Sunderland, Massachusetts, pp 251-265

Doty RW (1983) Nongeniculate afferents to striate cortex in macaques. J Comp Neurol 218:159-173

Doty RW (1988) Schizophrenia: a disease of interhemispheric processes at forebrain and brain stem levels? Behav Brain Res (in press)

Doty RW, Giurgea C (1961) Conditioned reflexes established by coupling electrical excitation of two cortical areas. In: Fessard A, Gerard RW, Konorski J, Delafresnaye JF (eds) Brain mechanisms and learning. Blackwell, Oxford, pp 133-151

Doty RW, Negra N, Yamaga K (1973) The unilateral engram. Acta Neurobiol (Warsaw) 33:711-728

Downer JDdeC (1961) Changes in visual gnostic functions and emotional behavior following unilateral temporal pole damage in the "split-brain" monkey. Nature 191: 50-51

Driscoll P, Claustre Y, Fage D, Scatton B (1987) Recent findings in central dopaminergic and cholinergic neurotransmission of Roman high- and low-avoidance (RHA/VERH and RLA/VERH) rats. Behav Brain Res 26: 213

Eckenrode TC, Barr GA, Battisti WP, Murray M (1987) Acetylcholine in the interpeduncular nucleus of the rat: normal distribution and effects of deafferentation. Brain Res 418:273-286

Fernandez de Molina A, Hunsperger RW (1962) Organization of the subcortical system governing defence and flight reactions in the cat. J Physiol (Lond) 160:200-213

Fonberg E (1972) Control of emotional behaviour through the hypothalamus and amygdaloid complex. In: Physiology, emotion and psychosomatic illness. Ciba Fndn Symp 8 (new series), Elsevier, Amsterdam, pp 131-161

Fonberg E, Kostarczyk E (1980) Motivational role of social reinforcement in dog-man relations. Acta Neurobiol Exp 40:117-136

Fox CA, Fisher RR, Desalva SJ (1949) The distribution of the anterior commissure in the monkey (Macaca mulatta). J Comp Neurol 89:245-278

Fox NA, Davidson RJ (l986) Taste-elicited changes in facial signs of emotions and the asymmetry of brain electrical activity in human newborns. Neuropsychologia 24:417-422

Freedman DG, Freedman NC (1969) Behavioural differences between Chinese-American and European-American newborns. Nature (London) 224:1227

Frijda NH (1986) The emotions. Cambridge Univ Press, Cambridge, 544 pp

Gainotti G (1987) Disorders of emotional behaviour and of autonomic arousal resulting from unilateral brain damage. In: Ottoson D (ed) Duality and unity of the brain; unified functioning and specialisation of the hemispheres. Plenum, New York, pp 161-179

Gloor P, Olivier A, Quesney LF, Andermann F, Horowitz S (1982) The role of the limbic system in experiential phenomena of temporal lobe epilepsy. Ann Neurol 12: 129-144

Goddard GV (1972) Long term alteration following amygdaloid stimulation. In: Eleftheriou BE (ed) The neurobiology of the amygdala. Plenum, New York, pp 581-596

Goodall J (1986) The chimpanzees of Gombe. Harvard Univ Press, Cambridge, Massachusetts

Gott, PS, Hughes EC, Whipple K (1984) Voluntary control of two lateralized conscious states: validation by electrical and behavioral studies. Neuropsychologia 22: 65-72

Grill HJ, Norgren R (1978) Neurological tests and behavioral deficits in chronic thalamic and chronic decerebrate rats. Brain Res 143: 299-312

Groenewegen HJ, Ahlenius S, Haber SN, Kowall NW, Nauta WJH (1986) Cytoarchitecture, fiber connections, and some histochemical aspects of the interpeduncular nucleus in the rat. J Comp Neurol 249: 65-102

Gurusinghe CJ, Zappia JV, Ehrlich D (1986) The influence of testosterone on the sex-dependent structural asymmetry of the medial habenular nucleus in the chicken. J Comp Neurol 253:153-162

Halgren E, Walter RD, Cherlow DG, Crandall PH (1978) Mental phenomena evoked by electrical stimulation of the human hippocampal formation and amygdala. Brain 101:83-117

Hall E (1972) Some aspects of the structural organization of the amygdala. In: Eleftheriou BE (ed) The neurobiology of the amygdala. Plenum, New York, pp 95-121

Hall, KRL, DeVore I (1965) Baboon social behavior. In: DeVore (ed) Primate behavior. Holt Rinehart Winston, New York, pp 53-110

Hara K, Myers RE (1973) Role of forebrain structures in emotional expression in opposum. Brain Res 52:131-144

Herbuté S, Peczély P, Astier H, Baylé JD (1977) Effects of bilateral lesions of the nuclei habenulae on plasma thyroxine levels in Japanese quail. Experientia 33: 912-913

Hilton SM, Zbrozyna AW (1963) Amygdaloid region for defence reactions and its efferent pathway to the brain stem. J Physiol (Lond) 165: 160-173

Horel JA, Keating EG (1972) Recovery from a partial Klüver-Bucy syndrome in the monkey produced by disconnection. J Comp Physiol Psychol 79: 105-114

Horel JA, Keating EG, Misatone LJ (1975) Partial Klüver-Bucy syndrome produced by destroying temporal neocortex or amygdala. Brain Res 94: 347-359

Innocenti GM (1986) General organization of callosal connections in the cerebral cortex. In: Jones EG, Peters A (eds) Cerebral cortex, vol 5: Sensory-motor areas and aspects of cortical connectivity. Plenum, New York, 291-353

Iwai E, Yukie M (1987) Amygdalofugal and amygdalopetal connections with modality-specific visual cortical areas in macaques (Macaca fuscata, M. Mulatta, and M. fascicularis). J Comp Neurol 261:362-387

Jacobs MS, Morgane PJ, McFarland WL (1971) The anatomy of the brain of the bottlenose dolphin (Tursiops truncatus). Rhinic lobe (rhinencephalon) J Comp Neurol 141: 205-271

Jones EG (1985) The thalamus. Plenum, New York, 744-757

Joseph R, Gallagher RE, Holloway W, Kahn J (1984) Two brains, one child: interhemispheric information transfer deficits and confabulatory responding in children aged 4, 7, 10. Cortex 20: 317-33l

Karli P, Vergnes M, Eclancher F, Schmitt P, Chaurand JP (1972) Role of the amygdala in the control of "mouse-killing" behavior in the rat. In: Eleftheriou BE (ed) The neurobiology of the amygdala. Plenum, New York, 553-580

Keating EG, Kormann LA, Horel JA (1970) The behavioral effects of stimulating and ablating the reptilian amygdala (Caiman sklerops). Physiol Behav 5:55-59

Kendrick KM, Baldwin BA (1987) Cells in temporal cortex of conscious sheep can respond preferentially to the sight of faces. Science 236:448-450

LeDoux JE (1987) Emotion. In: Mountcastle VB, Plum F, Geiger SR (eds) Handbook of physiology, Sect 1: The nervous system, Vol V, Higher functions of the brain, Part 1. American Physiological Soc, Bethesda, Maryland, 419-459

Leonard CM, Rolls ET, Wilson FAW, Baylis GC (1985) Neurons in the amygdala of the monkey with responses selective for faces. Behav Brain Res 15:159-176

Levitt P (1984) A monoclonal antibody to limbic system neurons. Science 223: 200-301

Leyhausen P (1973) Verhaltensstudien an Katzen. Paul Parey, Berlin,

Lilly R, Cummings JL, Benson F, Frankel M (1983) The human Klüver-Bucy syndrome. Neurology 33: 1141-1145

MacLean PD (1952) Some psychiatric implications of physiological studies of frontotemporal portion of limbic system (visceral brain). EEG Clin Neurophysiol 4:407-418

Mandell AJ, Knapp S (1979) Asymmetry and mood, emergent properties of serotonin regulation. A proposed mechanism of action of lithium. Arch Gen Psychiatry 36: 909-916

Martin JB, Reichlin S (1987) Clinical neuroendocrinology. FA Davis Co, Philadelphia, pp 111-122

McDonald AJ (1982) Neurons of the lateral and basolateral amygdaloid nuclei: a Golgi study in the rat. J Comp Neurol 212: 293-312

McGlone J (1980) Sex diferences in human brain asymmetry: a critical survey. Behav Brain Sci 3: 215-263

Millhouse OE, DeOlmos J (1983) Neuronal configurations in lateral and basolateral amygdala. Neuroscience l0: 1269-1300

Mizuno N, Uchida K, Nomura S, Nakamura Y, Sugimoto T, Uemura-Sumi M (1981) Extrageniculate projections to the visual cortex in the macaque monkey: an HRP study. Brain Res 212:454-459

Moller M, Korf H-W (1983) Central innervation of the pineal organ of the Mongolian gerbil. Cell Tissue Res 230: 259-272

Motohashi N, MacKenzie ET, Scatton B (1986) Functional mapping of the effects of lesions of the habenular nuclei and their afferents in the rat. Brain Res 397: 265-278

Muller RU, Kubie JL, Ranck JB Jr (1987) Spatial firing patterns of hippocampal complex-spike cells in a fixed environment. J Neurosci 7: 1935-1950

Murphy MR, MacLean PD, Hamilton SC (1981) Species typical behavior of hamsters deprived from birth of neocortex. Science 213: 459-461

Napier JR, Napier PH (1985) The natural history of the primates. MIT Press, Cambridge, Massachusetts

Nashold BS Jr, Wilson WP, Slaughter DG (1969) Sensation evoked by stimulation in the midbrain of man. J Neurosurg 30 :14-24

Nauta WJH (1961) Fibre degeneration following lesions of the amygdaloid complex in the monkey. J Anat 95:515-531

Nielson HC, Doty RW, Rutledge LT (1958) Motivational and perceptual aspects of subcortical stimulation in cats. Am J Physiol 194: 427-432

Nielson HC, McIver AH (1966) Cold stress and habenular lesion effects on rat behavior. J Applied Physiol 21: 655-660

Nishizuka M, Arai Y (1981) Organizational action of estrogen on synaptic pattern in the amygdala: implications for sexual differentiation of the brain. Brain Res 213: 422-426

Norris BK, Eichler VB (1975) The effect of ablation of the olfactory pits on the development of the habenular nuclei in Rana pipiens. J Embryol Exp Morph 345: 657-667

Northcutt RG (1981) Evolution of the telencephalon in nonmammals. Ann Rev Neurosci 4:301-350

Olds, ME, Olds J (1963) Approach-avoidance analysis of rat diencephalon. J Comp Neurol 120: 259-295

Olton DS, Becker JT, Handelmann GE (1979) Hippocampus, space, and memory. Behav Brain Sci 2313-365

Ono T, Fukuda M, Nishino H, Sasaki K, Muramoto K-I (1983) Amygdaloid neuronal responses to complex visual stimuli an in operant feeding situation in the monkey. Brain Res Bull 11: 515-518

Otto MW, Yeo RA, Dougher MJ (1987) Right hemisphere involvement in depression: toward a neuropsychological theory of negative affective experiences. Biol Psychiatry 22: 1201-1215

Overman WH, Jr, Doty RW (1979) Disturbance of delayed match-to-sample in macaques by tetanization of anterior commissure versus limbic system or basal ganglia. Exp Brain Res 37: 511-524

Pagni CA, Maroserro F (1965) Some observations on the human rhinencephalon - a stereo-encephalographec study. EEG Clin Neurophysiol 18: 260-271

Pandya DN, Seltzer B (1986) The topographe of commissural fibers. In: Leporé F, Ptito M, Jasper HH (eds) Two hemispheres - one brain, functions of the corpus callosum Alan R Liss, New York, 47-73

Panksepp J (1982) Toward a general psychobiological theory of emotions. Behav Brain Sci 5:407-467 Papez JW (1937) A proposed mechanism of emotion. Arch Neurol Psychiat (Chicago) 38 :725-743

Parent A (1986) Comparative neurobiology of the basal ganglia. Wiley, New York,

Pearson RCA, Sofroniew MV, Powell, TPS (1985) Hypertrophy of cholinergic

neurones of the rat basal nucleus following section of the corpus callosum. Brain Res 338 :337-340
Poblete R, Ruben RJ, Walker AE (1959) Propagation of after-discharge between temporal lobes. J Neurophysiol 22 :538-553
Poeck K (1969) Pathophysiology of emotional disorders associated with brain damage. In: Vinken PJ, Bruyn GW, Critchley M, Frederiks JAM (eds) Handbook of clinical neurology. vol 3, Disorders of higher nervous activity. Elsevier, New York, 343-367
Price JL (1986) Subcortical projections from the amygdaloid complex. In Schwarcz R, Ben-Ari Y (eds) Excitatory amino acids and epilepsy. Advances Exp Med Biol, Vol 203, Plenum, New York, 19-33
Raisman G, Field PM (1972) Functional implications of a quantitative ultrastructural analysis of synapses in the preoptic area and ventromedial nucleus. In: Eleftheriou BE (ed) The neurobiology of the amygdala. Plenum, New York, 83-94
Rausch LJ, Long CJ (1971) Habenular nuclei: a crucial link between olfactory and motor systems. Brain Res 29:146-150
Reiman EM, Raichle ME, Robins E, Butler FK, Herscovitch P, Fox P, Perlmutter J (1986) The application of positron emission tomography to the study of panic disorder. Am J Psychiatry 143: 469-477
Reinert H (1964) Defence reaction from the habenular nuclei, stria medullaris and fasciculus retroflexus. J Physiol (Lond) 170 :28P-29P
Rioch DMcK, Brenner C (1938) Experiments on the corpus striatum and rhinencephalon. J Comp Neurol 68: 491-507
Roberts E (1974) A model of the vertebrate nervous system. In: Myers RD, Drucker-Colin RR (eds) Neurohumoral coding of brain function. Plenum, New York, 419-449
Ronnekleiv OK, Kelly MJ, Wuttke W (1980) Single unit recordings in the rat pineal gland: evidence for habenulo-pineal neural connections. Exp Brain Res 39: 187-192
Rosene DL, Van Hoesen GW (1987) The hippocampal formation of the primate brain; a review of some comparative aspects of cytoarchitecture and connections. In: Jones EG, Peters A (eds) Cerebral cortex, Vol 6. Plenum, New York, 345-456
Russchen FT (1986) Cortical and subcortical afferents of the amygdaloid complex. In: Schwarcz R, Ben-Ari Y (eds) Excitatory amino acids and epilepsy. Advances Exp Med Biol, Vol 203. Plenum, New York, 35-52
Schaltenbrand G, Spuler H, Wahren W, Wilhelmi A (1973) Vegetative and emotional reactions during electrical stimulation of deep structures of the brain during stereotactic procedures. Z Neurol 205: 91-113
Schloeth R (1956) Zur Psychologie der Begegnung zwischen Tieren. Behaviour 10: l-80
Schreiner L, Kling A (1956) Rhinencephalon and behavior. Am J Physiol 184: 486-490
Sergent J (1986) Subcortical coordination of hemisphere activity in commissurotomized patients. Brain 109: 357-369
Sperry RW, Zaidel E, Zaidel D (1979) Self recognition and social awareness in the deconnected minor hemisphere. Neuropsychologia 17: 153-166
Steiner JE (1979) Human facial expressions in response to taste and smell

stimulation. In: Reese HW, Lipsitt LP (eds) Advances in child development and behavior. Academic Press, New York, 257-295
Sutherland RJ (1982) The dorsal diencephalic conduction system: a review of the anatomy and functions of the habenular complex. Neurosci Behav Rev 6: 1-13
Sutherland RJ, Nakajima S (1981) Self-stimulation of the habenular complex in the rat. J Comp Physiol Psychol 95: 781- 791
Swadlow HA, Geschwind N, Waxman SG (1979) Commissural transmission in humans. Science 204:530-531
Szentàgothai J, Flerkò B, Halàsz B (1962) Hypothalamic control of the anterior pituitary. Akadémia Kiadò, Budapest, 143-173
Tarr RS (1977) Role of the amygdala in the intraspecies aggressive behavior of the iguanid lizard, Sceloporus occidentalis. Physiol Behav 18: 1153-1158
Thatcher RW, Walker RA, Giudice S (1987) Human cerebral hemispheres develop at different rates and ages. Science 236: 1110-1113
Thornton EW, Evans JC (1982) The role of the habenular nuclei in the selection of behavioral strategies. Physiol Psychol 10: 361-367
Tigges J, Tigges, M (1985) Subcortical sources of direct projections to visual cortex. In: Peters A, Jones EG (eds) Cerebral cortex, Vol 3, Visual cortex. Plenum, New York, 351-378
Tizabi Y, Thoa NB, Maengwyn-Davies GD, Kopin IJ, Jacobowitz DM (1979) Behavioral correlation of catecholamine concentration and turnover in discrete brain areas of three strains of mice. Brain Res 166: 199-205
Turner BH, Mishkin M, Knapp ME (1979) Distribution of the anterior commissure to the amygdaloid complex in the monkey. Brain Res 162: 331-337
Valenstein ES. (1973) Brain control, a critical examination of brain stimulation and psychosurgery. Wiley, New York, 209-232
Valenstein ES, Cox VC, Kakolewski JW (1970) Reexamination of the role of the hypothalamus in motivation. Psychol Rev 77: 16-31
Wada JA, Mizoguchi T, Komai S (1981) Cortical motor activation in amygdaloid kindling: observations in nonepileptic rhesus monkeys with anterior two-thirds callosal bisection. In: Wada JA (ed) Kindling 2. Raven, New York, 235-248
Wilson JR, Mitchell JC, VanHoesen GW (1972) Epithalamic and ventral tegmental contributions to avoidance behavior in rats. J Comp Physiol Psychol 78: 442-449
Wirtshafter D (1981) The role of the interpeduncular connections with the tegmentum in avoidance learning. Physiol Behav 26: 985-989
Woods JW (1956) "Taming" of the wild Norway rat by rhinencephalic lesions. Nature 178: 869
Wyrwicka W (1978) Imitation of mother's inappropriate food preference in weanling kittens. Pavlovian J Biol Sci 13: 55-72
Yakovlev PI, Lecours A-R (1967) The myelogenetic cycles of regional maturation of the brain. In: Minkowski A (ed) Regional development of the brain in early life. Blackwell, Oxford, 3-70
Yakovlev PI, Locke S (1961) Limbic nuclei of thalamus and connections of limbic cortex III: corticocortical connections of the anterior cingulate

gyrus, the cingulum, and the subcallosal bundle in the monkey. Arch Neurol 5: 364-700

Zola-Morgan S, Squire LR, Amaral DG (1986) Human amnesia and the medial temporal region: enduring memory impairment following bilateral lesion limited to field of the hippocampus. J Neurosci 6: 2950-2967

Neurochemistry of the Emotions: Behavioural and Physiological Correlates of Catecholaminergic Systems

G. Bernardi, P. Calabresi and N. B. Mercuri

Neurological Clinic, II University of Rome, Via O. Raimondo, 00173 Rome, Italy

INTRODUCTION

In his classic paper "What is emotion?" William James (1893) postulated that emotions are generated by peripheral stimulations, such as increased heart and breathing rates. This theory (James-Lange theory) has given rise to several studies with the aim of proving an increased peripheral activation in patients affected by psychiatric disorders. Some years later Cannon (1932), in contrast with the James-Lange theory, suggested that the emotions are generated by the "arousal" of the central nervous system (CNS) and the peripheral effects are only secondary events. Cannon's (1932) classic experiments confirmed that it is the activation of the brain rather than the viscera that is relevant to emotional "arousal". With visceral nerves severed, emotional behaviour remained; lesions of brain areas caused primitive emotional responses.

The recent progress concerning the knowledge of neuronal physiology leads to the idea of relating some emotional states to the "activation" of distinct neuronal systems. In these processes chemical transduction is involved and the catecholaminergic systems play a major role. The most important pathways augmenting brain activity are mediated by catecholamine neurotransmitters, noradrenaline (NA) and dopamine (DA).
Although the complex interdependance between neuronal regulators suggests caution in interpreting the function of these pathways in isolation, there are characteristic effects of altering NA and DA neurotransmission.

Pribram and McGuinnes (1975) discussed a range of evidence that supports the differentiation between "arousal," mediated by NA systems, and "activation," originating from DA systems. "Arousal" produces a phasic response to sensorial inputs and "activation" maintains readiness for action by activating the basal ganglia.

Both these systems operate synergistically with other neurotransmitter systems. NA functions are in a reciprocally balancing relation to serotonin in "arousal," while DA pathways combine with cholinergic influences on the basal ganglia to regulate "activation" (McGuinnes and Pribram 1980).

Experimental Brain Research Series 18

Among the catecholaminergic theories of the emotions, the theory of the locus coeruleus was the most widely accepted (Grant and Redmond 1981). The primary NA pathway, the dorsal tegmental bundle, originates in the pontine locus coeruleus and projects to the median forebrain bundle and limbic system, including the amygdala, hippocampus anterior olfactory cortex, and septum, as well as the thalamus and neocortex.

The widespread innervation of the brain areas is remarkable, originating from a relatively small number of NA cell bodies in the locus coeruleus. The role of NA neurotransmission in regulating the organism's general level of "arousal" is shown by variation in the activity of the NA pathway with variations in the animal's sleep-wake cycle. In addition, electrical stimulation of locus coeruleus produces fear in animals similar to the "panic attacks" in humans; on the contrary the surgical ablation of these areas abolishes anxiety (Redmond 1979).

The GABA-benzodiazepine complex is another receptor site implicated in the central control of mood changes. It has been hypothesized that benzodiazepine reduces anxious behaviour by increasing the GABA-mediated opening of chloride channels (Tallman and Gallager 1985).

A THEORY OF THE EMOTIONS AT THE CELLULAR LEVEL

Although great progress in the neuropsychopharmacological approaches to the emotional states has been made, there are still several open questions concerning the relationship between systemic and cellular physiology and mood changes. The difficulties in relating data obtained from single neurons to behavioural concepts such as "arousal" and "activation" originate from the complexity of the intracellular and intercellular transduction mechanisms. These difficulties arising from the acquisition of modern concepts in neuroscience may be summarized as follows:

Intercellular Circuits.

Several structures modulating emotional states possess a complex intercellular morphofunctional organization. For example, in the CA1 hippocampal area the synaptic inhibitory processes are mediated by two separate inhibitory systems: feedback and feedforward circuits. Each of these systems possess separate anatomical and functional ways of activating GABAergic neurons (Alger and Nicoll 1982).

Types of synapses.

Modern acquisitions from microscopy and physiology regarding the synaptic organization of the brain have been consistent with some of the classical notions, but inconsistent with others. Surprising were the findings of synapses between axons and dendrites, of reciprocal synapses and of a variety of functional properties of dendrites (Shepherd 1974). Some synaptic interactions operate mainly in the distal dendrites of the neuron and involve voltage-dependent conductances (i.e., calcium conductances) typical of these areas and absent in somatic regions (Mountacastle and Bloom 1986).

Electrical Synapses.

Recently it has been shown that in some brain areas neurons are electrically coupled and chemical transmission is not necessary for the transduction of signals from one neuron to another one. However, it has been demonstrated that neurotransmission may also modulate the electrical synapses by modifying the amplitude of the diameter of the channels in the electrical junction (Lasater and Dowling 1985). It has also been hypothesized that electrical transmission may play a role in the genesis of the paroxismal activity (field bursts) in some regions of the brain (hippocampus and cortex), (Snow and Dudek 1984).

Cotransmitters.

The potential variety of transmitter interactions is further multiplied by the possibility of the coexistence and the corelease of messengers within the same nerve terminals. The function of some secondary cotransmitters may be that of modulating the presynaptic release or the postsynaptic action of the primary cotransmitters. Examples of possible coexisting substances in the CNS are CCK in DA-containing cells and in fibers projecting from the substantia nigra pars compacta and the ventral tegmental area to forebrain structures (Mountcastle and Bloom 1986, Skirboll et al. 1981).

Multiple Receptors and Autoreceptors.

Amongst the neurotransmitters and neuromodulators involved in the neurochemistry of emotions, in the past years, many have been shown to have multiple receptors (Aghajanian and Vander-Maelen 1982; Madison and Nicoll 1986; Mountcastle and Bloom 1986,Williams et al. 1982). NA receptors are classified as alfa 1, alfa 2, beta 1, beta 2; opioid receptors as mu, delta, kappa, sigma and DA receptors as D1 and D2. Thus, the same neurotransmitter may induce distinct and even opposite physiological responses on the same neuron interacting with different receptors. There is also evidence that some neurotransmitters reduce their own release from nerve terminals. This process is thought to be mediated by the so-called "autoreceptors" (Aghajanian and Bunney 1977; Chesselet 1984). The pharmacological properties of autoreceptors appear to be very similar to those of D2 sites. These receptors are activated by DA released from presynaptic terminals and act on a feedback mechanism to regulate the output of the transmitter from DA terminals.

Different Neurotransmitters Acting on the Same Channels

It has been widely shown that some aminoacids with inhibitory neurotransmitter functions may act on the same channels (Mountcastle and Bloom 1986). For example, GABA and glycine inhibit central neurons by opening chloride channels. Recently, in the locus coeruleus it has been shown that NA, acting on alfa receptors, and opioids, hyperpolarize membrane potential by opening the same potassium conductance (Aghajanian and Vander-Maelen 1982; Williams et al. 1982). On the contrary, in hippocampus, NA, (acting on beta receptors), serotonin, and istamine reduce the same calcium-dependent potassium conductance (Madison and Nicoll 1986).

Voltage-dependent channels.

Traditionally, ion channels in nerve cell membranes have been classified into two types: those opened by the action of neurotransmitters on receptor molecules causing excitation or inhibition and those opened by changes in membrane potential. Recently, it has become increasingly clear that ion channels controlled by membrane voltage can also be affected by neurotransmitters (Calabresi and Misgeld 1985; Mountcastle and Bloom 1986; Nicoll 1982). When modulation occurs on the cell body the firing pattern of cells can be drastically changed; when voltage-sensitive ion channels in the presynaptic terminals are the target, the amount of neurotransmitter release can be modified. Since neuronal membrane possesses a variety of voltage-sensitive channels which can be influenced by neurotrasmitters, it is apparent that neurotransmitters instead of being just exciting or inhibiting, actually exert a much more complex modulation.

Second messengers.

Second-messenger systems play a major role in mediating neurotransmitter action of substances involved in the neurochemistry of the emotions (DA, NA, Ach). There are two major receptor-regulated second-messenger systems currently thought to be active in the brain. The best known second-messenger system involves receptor-mediated generation of cyclic AMP, which in turn activates specific phosphorilating enzymes protein kinases (Greengard 1976, 1980). The more recently discovered phosphoinositide cycle appears to constitute a comparably universal second-messenger system involving the mobilization of internal calcium (Worley et al. 1987). Although second messengers have been known to exist for several years, their molecular characterization is currently ongoing. Several studies indicate that important therapeutic agents may by-pass receptor recognition sites to interact directly with second-messenger systems (Worley et al. 1987).

Physiological and Behavioural Correlates of NA neurotransmission

Admitting that difficulties exist in understanding the biological transduction of nervous signals and thus in understanding the cellular basis of emotions does not mean that progress has not been made in comprehending the biology of emotions. The most concrete results obtained at present are those regarding the mechanisms mediated by catecholamines.

The NA responses have been studied in hippocampus by utilizing slice preparations and intracellular recordings from CA1 areas (Madison and Nicoll 1986). These studies showed that NA excites or inhibits these neurons by interacting with different receptors. Of particular significance in understanding the role of NA in hippocampus is the observation that NA produces an increase in the number of spikes fired in response to depolarizing stimuli, whereas under control conditions the spike frequency of CA1 neurons rapidly diminishes with sustained depolarization ("accomodation"). In presence of NA, prolonged trains of action potentials are activated. This effect is believed to result from a depression of the late after-hyperpolarization. This action is mediated by beta receptors and involves cyclic AMP. In some cases, NA can also induce a weak membrane hyperpolarization mediated by alpha receptors. This evidence has suggested an alternative mechanism to explain the "signal to noise" enhancing effect of NA. Hyperpolarization would tend to decrease background firing induced by weak synaptic inputs. On the other hand, blockade of the spike after-hyperpolarization would selectively enhance repetitive spiking in response to strongly depolarizing synaptic inputs . In concert, these two effects would selectively favour transmission of activity elicited by strong stimuli and these data may provide a useful model of "arousal". In this sense, "arousal" would be considered as a behavioural and emotional state providing the organism with the ability to select specific stimuli within CNS.
It has been shown that the NA neurons of the locus coeruleus are strongly inhibited by alpha 2 receptors (Aghajanian and Vander-Maelen 1982) stimulating agents. Alpha 2 receptors also seem to be involved in the psychoparmacological mechanisms underlying panic attacks in humans (Gold et al. 1979). In opiate-addicted patients, the sudden interruption of drug intake produces anxious states comparable to those observed during panic attacks; clonidine, an alpha 2 agonist, is able, at least in part, to reverse such a symptomatology (Gold et al. 1979). These clinical observations, together with the experimental evidence showing that opioids (Williams et al 1982) and alpha 2 agonists activate different receptors present on the membrano of tho same neurons of locus coeruleus and inhibit these cells by opening the potassium channels, lead us to speculate that clonidine is able to block panic attacks, caused by the withdrawal syndrome, because it mimics the opioid-induced inhibition of locus coeruleus neurons.

Physiological and Behavioural Correlates of DA Neurotransmission

Pribram and McGuiness (1975) suggested that the activation is regulated primarily by the DA system of the basal ganglia. The activation is integral to motor operations and supports postural readiness and motivationally directed action. In contrast to "arousal" for which brain activity is entrained to the stimulus information, the attention directed by the "activation system" is more internally controlled by the extrapiramidal system.

The primary dopaminergic pathway in the brain originates from the substantia nigra in brainstem and innervates the caudate nucleus and the putamen, the neostriatum of the basal ganglia. A related DA pathway is the mesocortical or mesolimbic system, with cell bodies in the ventral nucleus and connections to limbic structures, including the olfactory tubercule, nucleus accumbens, central amygdaloid nucleus, and lateral septal nuclei. DA receptors have been observed in discrete areas of frontal cortex and in the entorinal cortex. Although an excitatory influence on motor behaviour follows an increase in the activity of dopami nergi c system, the primary effect of DA on target neurons appears to be inhibitory (Bernardi at al. 1978, 1984). These apparently paradoxical pieces of evidence can be explained by considering the circuits within the basal ganglia. The efferents from the striatum, in fact, regulate the functioning of the globus pallidus, which, in turn, exerts an inhibitory control over thalamic motor centers.

It repeatedly has been demonstrated that surgical or chemical manipulations which decrease activity of DA system of adult animals produce hypoactivity (for a rewiev see Schutz 1982). Thus, locomotor activity is reduced by DA blocking drugs, by 6-OHDA or electrolitic lesions of nigra or drugs such as reserpine and alpha-methil-para-tyrosine that deplete catecholamines. Drugs that enhance transmission at DA synapses, on the other hand, increase locomotor activity or produce stereotypies depending on the dose; these stimulating effects are produced by d-amphetamine, apomorphine, cocaine and other DA agonists.

In our laboratory, in the past years, we studied the action of DA on rat neostriatal neurons intracellularly recorded in vivo (Bernardi et al.1978, 1984; Mercuri at al.1985). DA inhibited spontaneous and evoked action potentials and depressed synaptic excitation. Recently, by utilizing in vitro slices preparations from neostriatum, we studied the ionic and the pharmacological mechanisms underlying the inhibitory action of DA in these structures (Calabresi et al.1987 a,b). In vitro, the inhibitory effect caused by micromolar concentrations of DA is mimicked by SKF 38393, a D1 agonist, and by 8-bromo-cAMP, a stable analogue of cAMP, but not by bromocriptine, lysuride and quinpirole which are D2 agonists. The inhibitory effects of DA are antagonized by SCH 23390, a D1 antagonist, but not by sulpiride, a D2 antagonist. The features of the DA-induced inhibition (persistence in absence of calcium, voltage-dependent behaviour, decrease of the depolarizing rectification of the neuron) suggest a postsynaptic site of action for the D1 mediated inhibition. Thus, D1 receptors, by reducing voltage- dependent inward conductance, inhibit the neuron only when it is depolarized by an excitatory input. On the contrary, D2 receptors would mediate inhibition of a presynaptic release from selective excitatory terminals (glutamatergic and cholinergic). It is interesting to note that when the neostriatal D2 receptors become supersensitive following DA-depleting procedures they exert both a pre and a postsynaptic inhibitory action.

In conclusion, it is possible to hypothesize that the function of DA in the neostriatum is to filter the inputs to this structure by utilizing a double receptorial mechanism: D1 receptors dump the response to any depolarizing signal and D2 receptors modulate selectively presynaptic release from specific projections. This hypothesis fits with the model of Pribram in which the DA system plays an activating role and maintains readiness for action providing the motor substrate for the emotional states and tho mood changes. The variety of subcellular, cellular and intercellular mechanisms operating in the transduction of the nervous signals did not completely impair the understanding of the catecholaminergic involvement in mood changes. The technical and theoretical acquisitions of future years will be crucial in further developing experimental approaches to the neurochemistry of the emotions.

REFERENCES

Aghajanian G K, Bunney B S (1977) Dopamine "autoreceptors": pharmacological characterization by microiontophoretic single cell recording studies. Naunyn Schmiedebergs Arch Pharmachol 297: 1-7

Aghajanian G K, Vander-Maelen C P (1982) Alpha 2-adrenoreceptors-mediated hyperpolarization of locus coeruleus neurons: intracellular studies in vivo. Science 215: 1394-1396

Alger B E, Nicoll R A(1982) Feed-forward dendritic inhibition studied in vitro. J Physiol (Lond) 328: 105-123

Bernardi G, Marciani M G, Morocutti C, Pavone F, Stanzione P (1978) The action of dopamine on rat caudate neurons intracellularly recorded. Neurosci Lett 8: 235-240

Bernardi G, Calabresi P, Mercuri N, Stanzione P (1984) Dopamine decreases the amplitude of excitatory post-synaptic potentials in rat striatal neurons. In: (eds) McKenzie M S at al. Basal ganglia :structure and function. Plenum, New York, 161-171

Calabresi P, Misgeld U (1985) Muscarinic modulation of Ca-mediated plateau spikes in rat neostriatal slices. Neurosci Lett 22: 280

Calabresi P, Mercuri N, Stanzione P, Bernardi G (1987a) Role of D1 and D2 dopamine receptors in the mammalian striatum: electrophysiological studies and functional implications. In:(ed) Carpontor M Basal ganglia II: structure and function. Plenum, New York, in press

Calabresi P, Mercuri N,Stanzione P, Stefani A, Bernardi G(1987b) Intracellular studies on the dopamine-induced firing inhibition of neostriatal neurons in vitro: evidence fo a D1 receptor involvement. Neuroscience 20: 757-771

Cannon W B (1932) The wisdom of the body. Norton, New York

Chesselet M F (1984) Presynaptic regulation of neurotransmitter release in rat brain. Neuroscience 2: 347-375

Gold M S Redmond D E Jr, Kleber H D (1979) Noradrenergic hyperactivity in opiate withdrawal supported by clonidine reversal of opiate withdrawal. Am J Psychiatry 136: 100-102

Grant S, Redmond D E Jr (1981) The neuroanatomy and pharmacology of the nucleus coeruleus. In: Fielding S (ed) The Psychopharmacology of clonidine, Liss, New York

Greengard P (1976) Possible role for cyclic nucleotides and phosphorylated membrane proteins in post-synaptic actions of neurotransmitters. Nature 234: 100-104
Greengard P (1980) Intracellular signals in the brain. Harvey Lect 75: 277-331
James W (1893) The principles of psychology vols. 1,2 Holt, New York
Lasater E M, Dowling J E (1985) Dopamine decreases conductance of the electrical junctions between cultured retinal horizontal cells. Proc Natl Acad Sci USA 82: 3025-3029
Madison D v, Nicoll R A (1986) Actions of noradrenaline recorded intracellularly in rat hippocampal CA1 pyramidal neurons in vitro. J Physiol (Lond) 372: 221-244
Mc Guinness D, Pribram K (1980) The neuropsychology of attention: emotional and motivational controls. In : Wittrock M C (ed) The brain and psychology Academic, New York
Mercuri N, Bernardi G, Calabresi P, Cotugno A, Levi G, Stanzione P (1985) Dopamine decreases cell excitability in rat striatal neurons by pre and post-synaptic mechanisms. Brain Res 358: 110-121
Mountcastle V B, Bloom F E (1986) The nervous system, sect. 1 Handbook of Physiology. American Physiological Society, Bethesda Maryland
Nicoll R A (1982) Neurotransmitters can say more than just "yes" or "no". Trends Neurosci 378: 369-374
Pribram K H, Mc Guinness D (1975) Arousal, activation and effort in the control of attention. Psychol Rev 82:116-149
Redmond D E (1979) New and old evidence for the involvement of a brain norepinephrine system in anxiety. In: Fann W E et al. (ed) Phenomenology and treatment of anxiety SP Medical and Scientific Books, New York
Schultz W (1982) Depletion of dopamine in the striatum as an experimental model of parkinsonism: direct effects and adaptative mechanisms. Prog Neurobiol 18:121-166
Shepherd G M (1974) The synaptic organization of the brain. Oxford University Press, Oxford.
Skirboll L R, Grace A A, Hommer D W, Rehfelì J, Goldstein M, Hokfelt T, Bunney B S (1981) Peptide-monoamine coexistence:studies of the actions of cholecystokinin-like peptide on the electrical activity of midbrain dopamine neurons. Neuroscience 6: 2111-2124
Snow R W, Dudek F E (1984) Electrical fields directly contribute to action potentials synchronization during convulsant-induced epileptiform bursts. Brain Res 323:114-118
Tallman J F, Gallager D W (1985) The Gaba-ergic system: a locus for benzodiazepine actions. Ann Rev Neurosci 8: 21-44
Williams J, Egan T M, North R A (1982) Enkephalines open potassium channels on mammalian central neurons. Nature 299: 74-77
Worley P F, Baraban J M, Snyder S H (1987) Beyond receptors: multiple second-messenger systems in brain. Ann Neurol 21: 217-229

Facial Expressive and Psychophysiological Correlates of Emotion

J. T. Lanzetta and G. J. McHugo
Department of Psychology, Dartmouth College, Hanover, New Hampshire 03755, USA

INTRODUCTION

In the past decade there has been a resurgence of interest in research on facial expressions of emotion. Several converging developments have contributed to the trend. The theoretical work of Tomkins (1962, 1963) and the seminal cross-cultural studies of Izard (1971) and Ekman (1972) demonstrated that fundamental issues, such as the relationships between facial expressions and emotional experience and the role played by the face in regulating social interaction, were theoretically tractable and capable of being empirically investigated in rigorous and creative ways. During this same period ethologists were providing rich descriptive data on primate expressive gestures and their role in establishing and controlling dominance hierarchies (Chance 1976; van Hooff 1973). They were also employing their methods to describe naturally occurring facial expressions (Redican 1982; van Hooff 1969) and were grappling with the problems of objectifying and standardizing a system for describing the repertoire of facial expressive gestures of primates and humans.

But, it was Ekman et al. (1971) who provided the biggest impetus for systematic work with their development of the Facial Affect Scoring Technique (FAST). This was followed by Ekman and Friesen's (1978) Facial Action Coding System (FACS) and Izard's (1979) Maximally Discriminative Movement Coding System (MAX). These formal measurement systems provided the first objective and componential analysis of facial expressions, permitting a more detailed study of the relationship between subjects' facial expressions and autonomic, subjective, and behavioral indices of emotion. However, these coding systems required highly trained coders and were time consuming and difficult to employ. They were also limited to the analysis of overt, detectable, facial actions. The measurement of muscle action potentials (MAPs) from surface electrodes overcame some of these disadvantages while introducing other measurement issues, for example, loss of detail, and problems of reconstructing a pattern from measurement of discrete muscles. Although many of these measurement problems remain unresolved, recent research employing one or another of the coding systems or electromyography (EMG) has provided a growing pool of data on the face and emotion.

Experimental Brain Research Series 18

A number of reviews have covered the literature on facial expressions (e.g., Ekman et al. 1972; Izard 1971). Ekman and Oster (1979) covered a wide range of topics and pointed to issues requiring further study. We will focus primarily on more recent studies and more especially studies done in our laboratories. We will also distinguish between studies that have examined spontaneous facial expressions in response to emotion-eliciting stimuli and those investigations directed to tests of the "facial feedback hypothesis." The latter have employed procedures for modifying or posing facial expressions in order to examine the effects of expressive responses on the nature and/or intensity of one's own emotions. The "spontaneous" studies are examined first, followed by consideration of the facial feedback investigations. Even with such restrictions, we can provide only a sample of the relevant work.

EMOTIONAL EXPERIENCE AND FACIAL EXPRESSIONS

Emotion theorists are not in accord regarding whether different emotions are characterized by distinctive autonomic nervous system (ANS) responses and/or expressive facial muscle patterned responses. Those purportedly following in the Jamesian tradition insist that ANS responses are not distinctive for different emotions, but vary only in magnitude. ANS activity is necessary, but context information provides the differential cues that determine which emotion is subjectively experienced (e.g., Schacter and Singer, 1962). Laird (1974), although endorsing the postulate of nondifferentiated ANS activity, suggests that facial expressions might differentiate among emotions and thus serve in place of context cues as a determinant of the subjective experience.

In contrast to this point of view, many investigators have argued that both ANS patterning and facial action patterns are distinctive for a number of fundamental emotions. The face especially is seen as serving intrapersonal and communicative functions, both of which require a highly differentiated, patterned, response system. To function as a signaling system, the face must provide discriminable cues for representative internal states, which are predictive of future actions. But does the face show distinctive expressions for each of a number of emotions, such as happiness, anger, fear, and so forth? And do the expressions vary systematically with differences in the intensity of the emotions?

A variety of approaches have been used to investigate the relationship between emotional states and facial expressions. The general paradigm employed by the majority of investigators is deceptively simple: while subjects are exposed to a stimulus that is presumed to evoke an emotional response, their facial expressive behavior is measured. If the facial expressions discriminate among emotion conditions or differentiate levels of intensity within an emotion, then a relationship is said to exist.

Difficulties with the paradigm arise at both the input and output sides. The favorite procedures for producing an emotion are posing (e.g., Ekman et al. 1983; Smith et al. 1986), imagery (e.g., Fridlund et al. 1984; Schwartz et al.1976; Smith et al.1986), slides (e.g., Cacioppo et al. 1986; Lanzetta et al. 1982); films (e.g., Ekman et al. 1980; McHugo 1983); direct sensory input such as electric shock, loud noise, or ice water (e.g., Lanzetta et al. 1976; Leventhal et al. 1979); and natural environmental stimuli such as phobic stimuli (e.g., Ohman et al. 1976). All of these procedures share some common problems. Among the array of possible exemplars, how do we select those that will evoke a particular emotion? What measures can be used as an independent check on the emotion-induction procedure? How can we determine whether a pure emotion or an emotional blend has been produced? The problems of operationalization are less acute, but not eliminated, when intensity and simple valence of the emotional arousal are of interest.

Investigators have used different approaches for evaluating their procedures for inducing an emotion, depending on the class of input stimuli employed. For example, with films, consensual validation of the arousal properties of the film, supplemented by subjects' ratings of their emotional response, is typical. With imagery procedures, subjects are assumed to "know" the qualities of the state, for example, sadness, which they are instructed to self-produce; here too subjective ratings are often employed as a manipulation check, but "demand" considerations limit their value. On occasion autonomic indicators are used to assess the effectiveness of a stimulus in eliciting an emotion. Since there is only fragmentary evidence that physiological "signatures" characterize particular emotions, their use is problematical. The appropriate methodological approach would be to use multiple, convergent measures; such an approach has seldom been employed.

On the output side, that is, the measurement of facial expression, there has been a rapid development of sophisticated and objective procedures. The earliest studies relied primarily on judges' ratings of still photos and, more rarely, dynamic displays. Such studies required subjects to give category judgments and were of little use in detailing specific facial actions. With the development of anatomically based or theory-based coding systems, attention could be directed to a much finer-grained analysis of expressive displays and to the specification of the pattern of muscle actions characteristic of a particular emotion. The more recent use of EMG techniques has sacrificed some of the analytic advantage of the visual coding systems, but EMG permits the measurement of covert expressive behavior and provides continuous, easy-to-record data.

Although we do not yet have unequivocal solutions for these methodological and measurement problems, such a variety of procedures have been used with similar results that a measure of confidence in the validity of the findings is warranted.

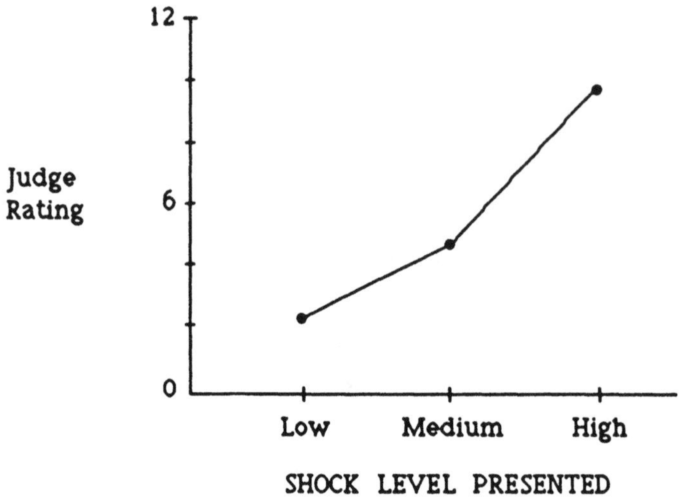

Fig. 1. Judges' rating of the intensity of shock experienced by subjects as a function of the level of shock delivered. (Adapted from Lanzetta et al. 1976)

FACIAL EXPRESSION AND THE INTENSITY AND VALENCE OF EMOTIONAL EXPERIENCE

Data relevant to this issue have come from studies using observer judgments, formal coding systems, and muscle action potentials. Lanzetta et al. (1976) obtained videotapes of subjects' faces while they were anticipating and receiving electric shocks of varying intensity. Judges, uninformed as to the shock levels used, were later asked to rate the intensity of the subjects' pain expressions. Figure 1 shows the relationship between judged intensity of the pain expressions and the physical magnitude of shock. The relationship is monotonically increasing, indicating that subjects' facial expressions covaried closely with the intensity of the pain experience.

Cacioppo et al. (1986) showed subjects slides of scenes that were mildly to moderately evocative of positive and negative affect. Facial EMG activity over the brow (corrugator supercilii), eye (orbicularis oculi), and cheek (zygomaticus major) regions differentiated the pleasantness and intensity of affective reactions to the visual stimuli, even though visual inspection of videotapes of subjects' faces indicated no apparent expressions of emotion. Cacioppo and Petty (1979) have also demonstrated that the mean amplitude of EMG activity recorded over selected facial muscle regions differentiated subjects exposed to a proattitudinal appeal from those exposed to a counterattitudinal appeal, and Cacioppo et al. (1984) found that localized changes in EMG differentated trials on which subjects imagined reading agreeable vs disagreeable editorials.

Facial EMG has also been shown to discriminate between the unconditioned responses of human subjects to reward, in the form of money, and punishment,

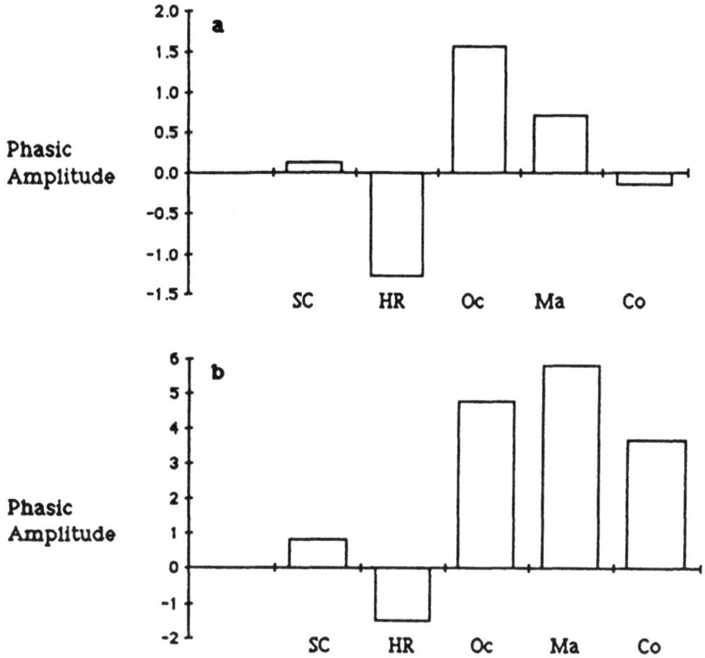

Fig. 2a, b Unconditioned emotional responses of observance averaged across all aquisition trials. a) Reward, b) Shock; SC, skin conductance (μmhos$^{.5}$); HR, heart rate (bpm); Oc, orbicularis oculi; Ma, masseter; Co, corrugator ($\mu V^{.5}$). (Adapted from Englis et al. 1982)

in the form of mild electric shock (Englis et al. 1982). There is an unambiguous subjective distinction between the two emotional responses, but because both outcomes are salient and arousing, psychophysiological differentiation through autonomic indicators may be weak or absent. On the other hand, facial expressions of emotion are expected to differ as a consequence of reward or punishment.

Figure 2 shows the average phasic response immediately after receiving reward (in the top panel) or punishment (in the bottom panel). The psychophysiological variables are skin conductance, heart rate, and EMG from three facial regions: obicularis oculi, which closes the eye; masseter, which clenches the jaw; and corrugator supercilii, which knits the brow. All three muscles are components of the grimace, and all contribute significantly here to the differentiation of the emotional response to reward from that to punishment.

The advantage of this improved definition of the unconditioned, or instigated, emotional response is in comparisons afforded with collateral conditioned emotional responses. Rather than using an electric shock directly delivered to a subject, Vaughan and Lanzetta (1980) exposed subjects to a vicariously

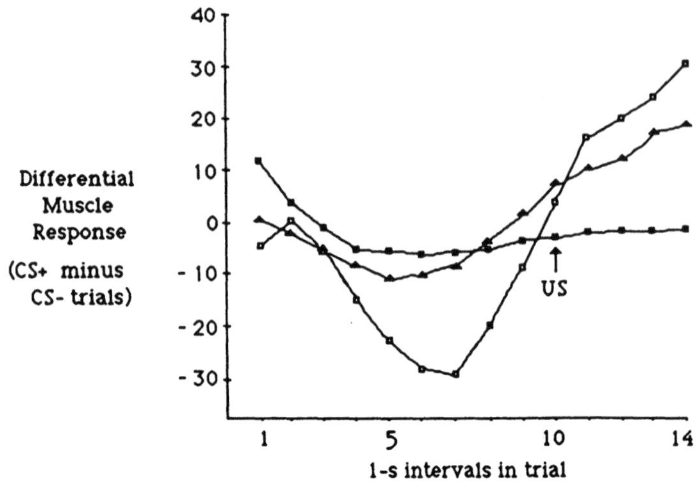

Fig. 3. Vicarious instigation and conditioning of orbicularis oculi (▲), masseter (◻), and medial frontalis (▪) muscles in experiment 2. The unconditioned stimulus (US) was presented 10 after trial onset. (Adapted from Vaughan and Lanzetta 1980)

induced aversive stimulus. Subjects saw a videotape of a "model" subject engaged in a learning task. Words in one of the two categories presented to the model were followed by grimaces to indicate the delivery of a shock punishment. Thus, the words served as conditioned emotional stimuli, and the model's facial response was the unconditioned stimulus for the observing subject. The results showed that the grimace of the model was adversive and instigated grimace-like facial responses in subjects, but the conditioned, that is, anticipatory, response to the model's expression of pain was one of inhibition of the facial muscles involved.

Figure 3 shows the time course of this difference between the anticipatory and the instigated responses for three facial muscles: orbicularis oculi, masseter, and medial frontalis, which is a forehead muscle. Plotted are second-by-second differences in average muscle region response between trials where the model grimaced, that is, CS+ trials, and trials where no pain display was present, that is, CS- trials. A score above zero indicates greater excitation in observing subjects on trials where the model grimaced vs trials where no outcomes were presented to the model. During the period after the stimulus word (CS) appears and prior to the model's grimace, which begins during the 10th s, the masseter and obicularis oculi regions show initial inhibition followed by anticipatory excitation. Excitation is increased as the subject observes the model's grimace. The medial frontalis shows initial relative relaxation, but no anticipatory or instigated excitation; it did not differentiate aversive (CS+) trials from neutral (CS-) trials. This differentiation by select facial muscles of vicariously conditioned from vicariously instigated responses occurred in the absence of such oppostion in the skin conductance responses, which showed a phasic increase during both periods. In this case then, the facial EMG measures allowed the categorization of the vicarious responses: first an anticipatory, or conditioned, fear/anxiety response, followed by an instigated pain response. Further, we could distinguish emotional reactions from a general change in arousal/tension; the lack of response of the frontalis muscle clearly signifies that increases in general tension did not take place.

Facial muscle differentiation among emotional responses, in both within- and between-subject conditions, was accomplished in a study of empathic and counter-empathic emotional responses (Englis et al. 1982). In this experiment subjects believed, through videotape simulation, that they were playing a computer game with a person in another room. In the symmetry, or empathy, condition subjects received outcomes that were consistent with those to the model. That is, the symmetry group received cash rewards on trials where they viewed the smile of the model, and they received shock punishments on the same trials where the model grimaced. In the asymmetry, or counter-empathy, condition subjects were punished after the model smiled and they were rewarded after the model grimaced. Thus, a partner's smile and grimace expressive responses acquired either positive or negative valence depending on the conditioning history. The question of interest was whether subject's emotional response to the partner's expression would reflect properties of the stimulus alone or encompass the conditioning.

After an acquisition phase, half punishment and half reward to both the subject and the model, there followed an extinction, or vicarious, phase where the subject simply observed the model play alone against the computer. The conditioned stimulus, that is, the model's facial display, thus remains, but the subject neither responds to the game nor receives outcomes.

Figure 4 shows the crossover interactions obtained from this mixed design, where the responses to the grimace and the smile of the model are shown for both the symmetry and the asymmetry groups during extinction. The top panel shows average skin conductance responses. The model's grimace evoked arousal in the symmetry group, because this cue had previously been associated with shock to the subject. On the other hand, the asymmetry group showed arousal to the smile and no change to the grimace, indicating the success of conditioning the counter-empathic emotional response. The lower panel shows the phasic EMG responses from the orbicularis oculi region. Again, the conditioned responses to the shock and reward outcomes differ by experimental groups. The symmetry group showed more tension in the eye region during the model's grimace than during the model's smile, while the opposite was true for the asymmetry group. The emotional responses to the model's facial expressions were very similar to the subject's own unconditioned responses to actual reward and punishment, which were shown in Fig. 2. The empathic response to another's pain display thus resembles the observer's own response to real pain, whereas the counter-empathic response to another's pain resembles the observer's own response to reward. These

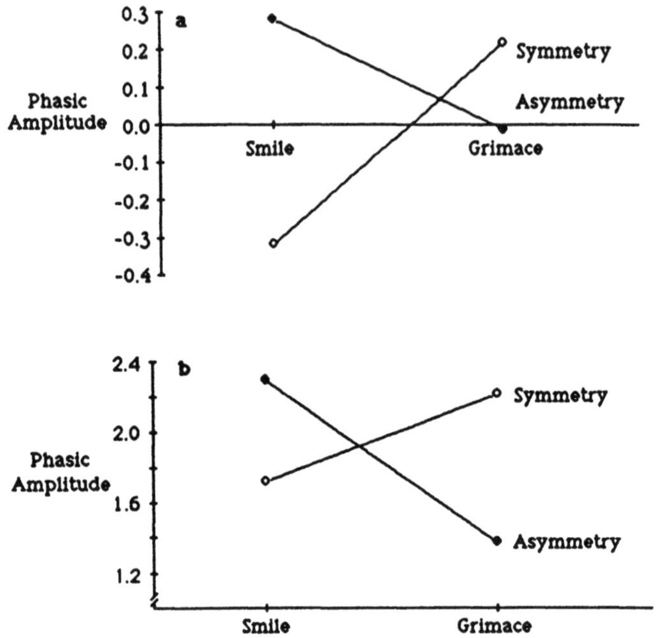

Fig. 4 a,b. Group by expression interactions persistent across all vicarious phase trials. a) Skin conductance (Δ μmhos$^{.5}$); b) Orbicularis oculi (Δ μV$^{.5}$). (Adapted from Englis et al. 1982)

EMG results converge with the electrodermal results to indicate that the facial expressions of another have positive or negative hedonic tone and thus elicit affect-specific facial expressive responses. The EMG results were found even though judges of the videotapes of the subjects' faces could not differentiate smile from grimace trials.

The evidence from the cited studies and others not discussed strongly supports the conclusion that facial expressions can differentiate both the valence and intensity of emotional arousal. Converging results have been obtained for a wide array of eliciting conditions including imagery, slides of evocative scenes, facial displays of conspecifics, direct adversive sensory inputs, and posing.

For valence of affect, judges have no difficulty in discriminating positive from negative emotions, although the specific facial expressive cues used by judges have not been identified. However, the consistent finding using electromyographic recordings is that EMG activity over the corrugator supercilii muscle (draws brows together and down) region is higher and that over the zygomaticus major muscle (draws the corner of the mouth up and back) region is lower during negative than positive emotions. On occasion investigators have found that the orbicularis oculi also differentiates positive from negative affective states much like that provided by the zygomatic muscle (Cacioppo et al. 1986). The latter is a somewhat anomolous finding, since when pain is the negative stimulus, the changes in corrugator and zygomatic regions are accompanied by increased activation in the orbicularis oculi muscle region (Englis et al. 1982).

For intensity of emotion, the data are less extensive, although consistent. Judges are able to discriminate levels of intensity of pain and happy expressions, and EMG studies indicate that the corrugator region is especially useful in discriminating levels of intensity of both pleasant and unpleasant stimuli. As Cacioppo et al. (1986, p 264) note, "...EMG activity over the muscle region responsible for lowering and drawing the eyebrows together (corrugator supercilii) tended to be higher for moderately than for mildly intense unpleasant affective reactions, whereas it was lower for moderately than for mildly intense pleasant reactions."

FACIAL EXPRESSIONS OF FUNDAMENTAL EMOTIONS

Facial expressions may differentiate positive from negative states and may discriminate levels of intensity of an emotional reaction, but provide poor differentiation among discrete emotions. Does the face show distinctive expressions for each of a number of emotions? If so, how many emotions have a distinctive facial expression?

Early research relied primarily on observer accuracy in judging posed emotions and focused on the question of universality. Are facial expressions of emotion culturally variable social signals or do members of different cultures show the same facial expressions when experiencing the same emotion? Ekman (1972), Ekman et al. (1969), and Izard (1971) used theory-based descriptive criteria to select photographs representative of a number of emotions, which were then shown to informants from a number of different cultures. Observers were asked to select the most appropriate descriptive adjective from a list of possible emotion labels. The majority of subjects in each of a number of literate cultures and in several preliterate cultures interpreted the facial expressions as conveying the same emotion. These authors concluded that in literate and preliterate cultures distinctive and shared facial expressions exist for anger, disgust, happiness, sadness (or distress), fear/surprise, and possibly for interest and shame. These distinctive expressions may be modified by cultural display rules, but the modifications are situation specific.

A limitation of these studies is that the facial expressions used as stimuli were not spontaneous, but were posed by subjects instructed to show a particular emotion. Buck et al. (1972, 1974) and Lanzetta and Kleck (1970) have used videotapes of subjects' faces as they observed emotionally evocative slides or while anticipating and receiving mild electric shock, respectively. Judges were able to apply correct labels to the subjects' facial expressions with better than chance accuracy.

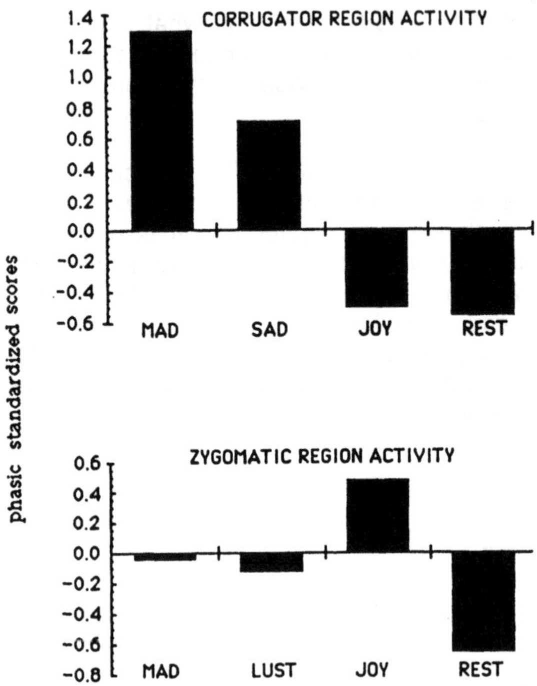

Fig. 5. Phasic standardized facial EMG responses (corrugator and zigomatic regions) during emotional imagery

Studies of emotional imagery have provided further evidence that, even in the absence of overt facial movement, different affective states can lead to specific changes in facial EMG activity. The research of Schwartz and his colleagues has been especially instructive in demonstrating patterned changes in EMG activity for different imagined emotional states (e.g., Fridlund et al. 1984; Schwartz et al., 1976). In our laboratory we have also examined EMG patterns for imagined and self-generated emotions. Subjects were asked to generate, with relevant imagery, feelings associated with several emotions, for example, sadness, anger, and happiness. Figure 5 shows the mean corrugator and zygomatic EMG activity for four of the self-generated emotions.

In a related experiment (Smith et al. 1986) subjects were instructed in imagery procedures and then asked, on two separate days, to imagine happy, angry, and sad events from their past. Two results are of interest. First, average EMG at the three facial muscle sites selected for measurement (corrugator supercilii, zygomaticus major, and depressor anguli oris) differentiated among the three emotional states, and second, the pattern of facial muscle activity did not differ over the two measurement occasions. The latter finding is of special interest, since data is lacking on the consistency of facial expressive responses to specific emotions.

Table 1. Correlations between facial EMG responses and sel-reports of emotion from evocative film segments that were significant beyond the 0.1 level (one tailed); (Adapted from McHugo 1983).

MASH		RED BALLON	
Zigomatic-Joyful	0.37(0.08)	Zigomatic-Amused	O.40(0.06)
Zigomatic-Sad	-0.35(0.09)	Zigomatic-Joyful	0.34(0.09)
Zigomatic-Angri	-0.34(0.10)	Zigomatic-Anxious	-0.50(0.02)
		Zigomatic-Fear	-0.37(0.08)

THE HUSTLER		THE INFORMER	
Zigomatic-Amused	0.55(0.02)	Zigomatic-Amused	0.75(0.001)
Zigomatic-Sad	-0.53(0.02)	Zigomatic-Scorn	-0.38(0.08)
Corrugator-Angri	0.36(0.09)	Zigomatic-Disgust	-0.40(0.07)
Corrugator-Anxious	-0.41(0.06)	Corrugator-Scorn	-0.42(0.06)
Corrugator-Amused	0.48(0.03)		

In this same experiment, subjects viewed six film segments that had been rated for emotional impact in an earlier study (McHugo et al. 1982). Of the six film segments viewed, two were mildly evocative, two were affectively positive, and two were affectively negative. Immediately after each segment, subjects rated the intensity of the emotions evoked by the films using the following descriptive adjectives: amused, joyful, sad, angry, tearful, anxious, disgusted, and contemptuous. A principle components factor analysis reduced the subjective evaluations to four factors, namely, joyful, sad, anxious, and disgusted.

Overall, both subjective reports and EMG measures discriminated among the film segments. Corrugator muscle activity was especially effective in differentiating the film segments from each other, including the two positive segments. To determine whether changes in facial muscle tension levels were related to the reported intensity of elicited emotions, bivariate correlations were calculated over subjects between the facial EMG and self-reported response for each evocative film segment. Table 1 presents these correlations. As the table shows, zygomatic activity is positively associated with positive affects and negatively associated with negative affect,

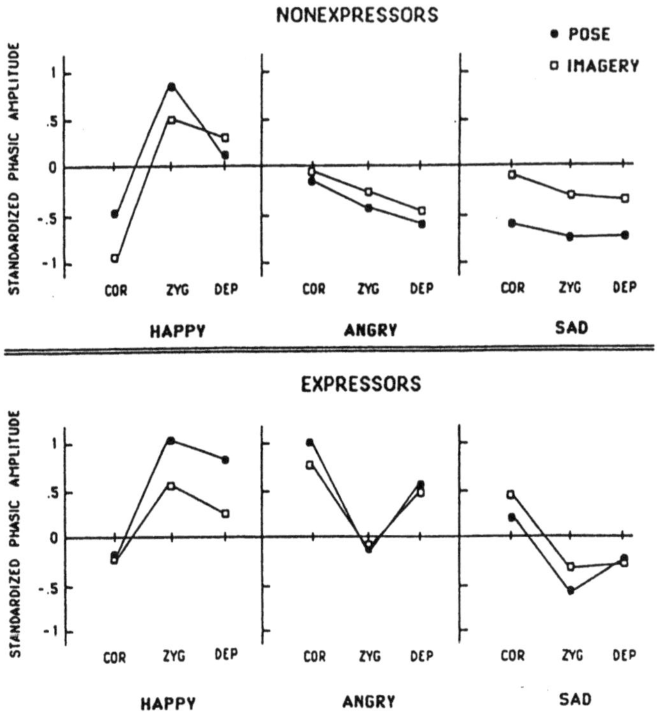

Fig. 6. Profiles of mean changes in facial muscle tension (Z scores) in the brow, cheek, and mouth regions for emotion conditions in the imagery and posing tasks for nonexpressive and expressive subjects. (Adapted from Smith et al. 1986)

whereas corrugator activity shows the opposite pattern of association (McHugo 1983).

The acute sensitivity of facial EMG to subtle changes in emotional state has proven especially effective in discriminating among individuals who differ in some important behavioral or attitudinal characteristics. Let us illustrate with two examples.

In a study of expressive behavior, subjects were induced to pose on two occasions happy, angry, and sad emotions. They were also instructed in imagery procedures and then asked again on two occasions, to imagine prior happy, angry, and sad events. We were primarily interested in whether posed and spontaneous expressions were consistent over time, and similar in pattern (if not in intensity). Secondarily, however, we were also interested in any possible differences that might exist between expressive and nonexpressive subjects. Since subjects were categorized on the basis of success in "posing," the primary comparison was among EMG responses in imagery. Figure 6 shows the results. Note that facial muscle patterns differ

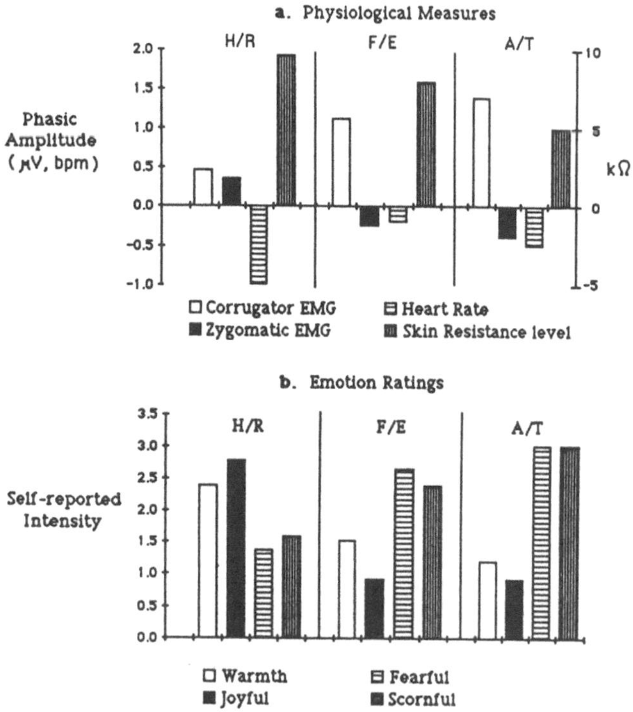

Fig. 7. Mean phasic psychophysiological responses during, and self-report profiles following, happines/reassurance (H/R), fear/evasion (F/E), and anger/threat (A/T) displays by President Reagan. (Adapted from Lanzetta et al. 1985)

for different emotions and that posed and imagery expressions are very much alike. Further, nonexpressors show a pattern much like that shown by expressors for happy conditions, but show almost no facial activity for sad and very little for angry imagery. Nonexpressors seem to be inhibiting expressions of negative affect (Smith et al.1986).

As a final illustration to show the use of such measures in more ecologically valid settings, let us turn to some recent studies of emotional responses to the expressive displays of political leaders.

In a series of experiments (see review by Lanzetta et al. 1985) subjects were exposed to videotaped excerpts of facial expressions of political leaders, for example, President Reagan. The expressive displays were selected to represent happiness/reassurance, anger/threat, and fear/evasion on the part of the political leader. Subjects' emotional responses during the viewing session were assessed by measures of autonomic activity (heart rate and skin conductance) and of facial expressive behavior (corrugator supercilii and zygomaticus major EMG); after the presentation of each facial display, verbal self-reports of emotion were obtained (McHugo et al. 1985).

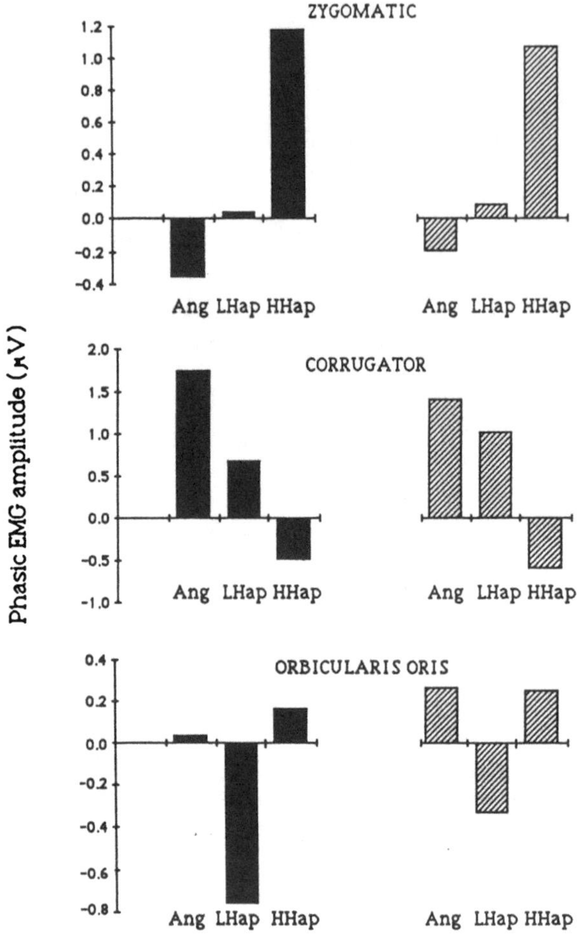

Fig. 8. Phasic facial EMG responses to President Reagan's (on the left) and Senator Hart's (on the right) displays of anger (Ang), mild happiness (LHap), and intense happiness (HHap). (Adapted from McHugo et al. 1988, to be published)

The results for the physiological measures and emotion ratings are shown in Fig. 7. The subjective reports indicate that subjects had distinctively different emotional reactions to the three expressive displays of President Reagan. Self-reported positive affect was higher and negative affect lower when viewing happy/reassurance displays, whereas the opposite pattern obtained when viewing the fear/evasion and anger/threat displays. The physiological measures also provided clear discrimination among the display conditions. Corrugator activity was progressively more intense and zygomatic activity less intense from the happy to fear to anger displays, whereas skin resistance level and heart rate deceleration decreased over the same stimulus conditions.

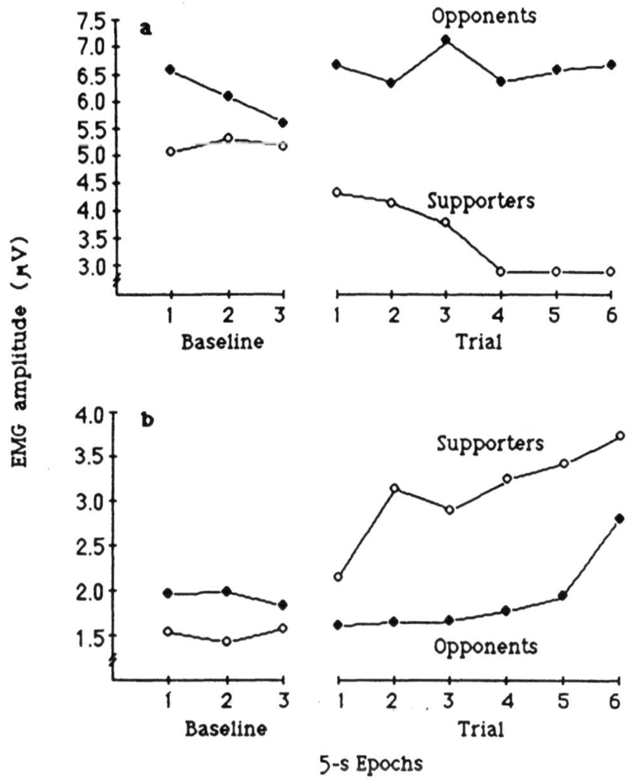

Fig. 9 a,b. Facial EMG (a) corrugator and B) zygomatic region) responses over time to President Reagan's intense happines displays as a function of prior attitude toward Reagan. (Adapted from McHugo et al. 1988, to be published)

An extension of this study used displays representing two levels of happiness in addition to anger/threat (McHugo et al. 1988, to be published.). Figure 8 shows the EMG activity for the corrugator, obicularis oris, and zygomatic muscle regions. The facial muscles provide sharp differentiation between happiness and anger. The figure also show that the expected differences in intensity of emotional responses of supporters and opponents of President Reagan are reflected in activity of the facial muscles. This finding is illustrated in a clearer fashion in Fig. 9 and 10, which show the facial EMG amplitudes over time while subjects viewed the displays. These are shown separately for supporters and opponents. The EMG levels appear to reflect the growth and decay of subjects' emotional reactions to President Reagan's displays and indicate that, although opponents and supporters have qualitatively similar emotional responses, they differ in both the latency and amplitude of their reactions.

It is evident from the sampling of studies reviewed that facial expressions vary as a function of emotional state. But the data do not permit one to

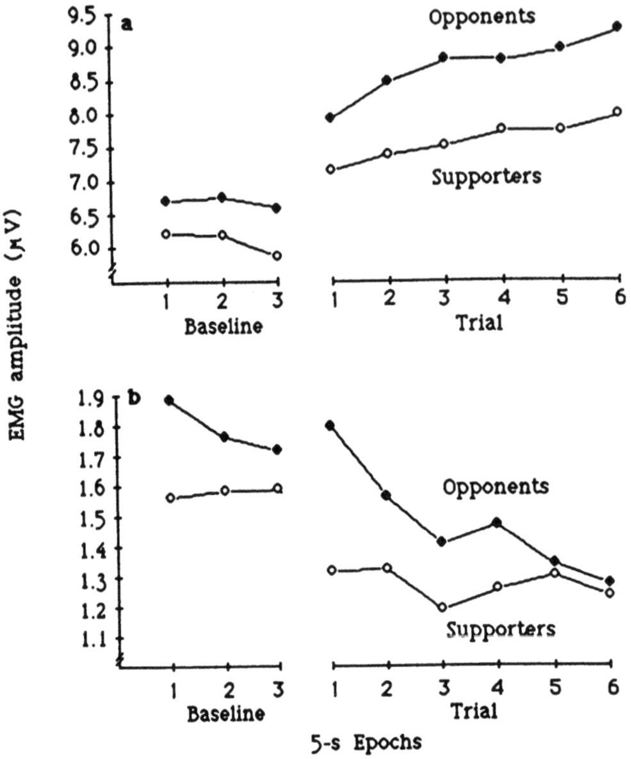

Fig. 10 a,b. Facial EMG (a) corrugator and B) zygomatic region) responses over time to President Reagan's anger displays as a function of prior attitude toward Reagan. (Adapted from McHugo et al. 1988, to be published)

specify the pattern of muscle actions characteristic of a particular emotion. Part of the problem of providing such specification is in establishing which emotion, if any, was experienced at the time of facial measurement. Self-reports are error-prone, autonomic indicators are themselves not well-specified enough to provide differentiating criteria, and vocal parameters are imprecise indicators of particular emotional states. In the absence of a single valid criterion, multiple convergent measures would provide a more reliable indication of the emotions experienced. Few studies have, in themselves, used multiple convergent measures, although the wide variety of approaches used by different investigators provides an approximation to this approach.

A further problem in determining the specific pattern of facial actions characteristic of an emotion is in the measurement procedures employed. Many of the eliciting conditions used by investigators, that is, imagery, films, and slides, do not always elicit observable facial actions. Thus, comprehensive coding systems, such as FACS or observer judgments, cannot be used effectively. Facial EMG has proven very useful in such situations, but few investigators have attempted to measure activity at more than two or three facial sites. Only a very restricted view of facial expressions can emerge from this approach. Without knowledge of eliciting stimuli, judges could not identify a specific emotion from the examination of a restricted pattern of EMG changes, although a trained observer could no doubt specify the hedonic valence of the experienced affect.

In the light of these methodological shortcomings what can we say about the relationship between facial expressions and emotion?

1. Facial expressions provide accurate information about the valence of experienced emotion. Pleasant emotions can be differentiated from unpleasant emotions using global observer judgments, visual coding systems, and facial EMG measures. The EMG studies suggest that the corrugator and zygomatic muscles provide the primary differentiating cues.
2. Facial expressions can provide information about the intensity of a pleasant or unpleasant state. The data for this assertion are limited and come primarily from EMG studies which suggest that corrugator and zygomatic MAPs monotonically increase with increases in the intensity of an emotion.
3. Facial expressions probably provide information about discrete emotion, but the data in support of this assertion are sparce. Posing studies using observer judgments are the richest source of data, but EMG studies indicate that some facial muscles can differentiate among several negative emotions.
4. Prior emotional dispositions of subjects to emotion-eliciting stimuli are reflected in their facial expressive displays. That is to say, subtle differences in the emotion elicited by a display are reflected in the latency and amplitude of the EMG responses in selective facial regions.
5. Individual differences in expressivity are not situation-specific, but may be emotion-specific. "Nonexpressive" subjects show little facial activity when experiencing negative emotions, but do not differ from expressive subjects when experiencing positive emotions.

CAUSAL LINKS BETWEEN FACIAL EXPRESSION AND EMOTION: THE FACIAL FEEDBACK HYPOTHESIS

In our selective review of the research on expressive behavior, we ignored the varying theoretical views regarding the link between the facial expression of an emotion and the subjective experience of that emotion. The empirical research shows that there is a relationship between the outward signs of an emotion and its experience, but is the relationship merely correlational or is it causal, and, if the latter, what is the direction of causation?

The "facial feedback hypothesis" actually refers to a variety of hypotheses, which have in common the prediction of a causal role for the face in the experience of emotion, but which differ in important ways in the specification of this role, as well as in theoretical underpinnings and proposed mechanisms. Broadly speaking, the facial feedback hypothesis states that if patterned feedback from facial expression contributes causally to the experience of an emotion, then modulation, or production, of the facial expression will influence emotional experience.

The historical antecedents of the facial feedback hypothesis begin with the work of several nineteenth century physiologists whose pioneering work on facial expression was integrated and popularized by Charles Darwin (1965/1872). In addition to explicating the role of facial expression in communicating emotions among individuals, Darwin proposed a causal role for facial expression in modulating intraindividual emotional experience:

"The free expression by outward signs of an emotion intensifies it. On the other hand, the repression, as far as this is possible, of all outward signs softens our emotions. He who gives way to violent gestures will increase his rage; he who does not control the signs of fear will experience fear in a greater degree; and he who remains passive when overwhelmed with grief loses his best chance of recovering elasticity of mind." (Darwin 1965, p 365).

The well-known "peripheralist" theory of William James also attributed a causal role to bodily action in the subjective definition of emotional states. This was wrongly considered by subsequent scholars to refer solely to autonomic activity, as they overlooked James' references to somatic activity:

"The feeling, in the coarser emotions, results from the bodily expression. Our natural way of thinking about these coarser emotions is that the mental perception of some fact excites the mental affection called the emotion, and that this latter state of mind gives rise to the bodily expression. My theory, on the contrary, is that the bodily changes follow directly the perception of the exciting fact, and that our feeling of the same changes as they occur is the emotion. Without the bodily states following on the perception, the latter would be purely cognitive, pale, colorless, destitute of emotional warmth. We might then see the bear and judge it best to run, receive the insult and deem it right to strike, but we should not actually feel afraid or angry." (James 1962 pp 377-378)

James referred to bodily expression and action, but he was less explicit than Darwin in implicating the face as a key element in emotional experience. Ironically, in the 1920s, while Cannon and others were leveling damaging criticism against the James-Lange theory of emotion, Floyd Allport (1924) proposed a theory in which the somatic system was again central. In discussing how emotions are differentiated beyond the simple pleasant-unpleasant dimension, Allport said:

"We propose that the differentiating factor arises from the stimulation of the proprioceptors in the muscles, tendons, and joints of the somatic part of the organism; and that afferent impulses from these somatic patterns of response add to the autonomic core of affectivity the characteristic sensory complexes by which one emotion is distinguished from another of the same affective class. Different, and somewhat antagonistic, somatic effector groups are brought into play according to whether the individual attacks or flees. The facial expressions as well as bodily movements are strongly differential. Return afferent impulses from these responses add in consciousness the distinguishing qualities which serve to differentiate the emotion of anger from that of fear." (Allport 1924, pp. 91-92)

Theory and research in human emotion lay relatively dormant during the dominant years of the behaviorist movement in psychology, until several important experiments, such as those of Ax (1953) and Schachter and Singer (1962), revitalized interest in the subject. At the same time, Tomkins (1962, 1963) published a landmark volume explicating a wide-ranging, but detailed, theory of emotion, which included a central role for facial expression in the experience of emotion. Concerning the differential involvement of the autonomic system and the face, Tomkins wrote:

"Just as the fingers respond both more rapidly with more precision and complexity than the grosser and slower moving arm to which they are attached, so the face expresses affect, both to others, and to the self, via feedback, which is more rapid and more complex than any stimulation of which the slower moving visceral organs are capable. It is the very gross and slower moving characteristic of the inner organ system which provides the counterpoint for the melody expressed by the facial solo." (Tomkins 1962, pp. 205-206)

Tomkins has modified his theory over the years (e.g., Tomkins 1980), but he still views feedback from facial expression as an important determinant of subjective experience.

Of course, not all theories of emotion endorse the view that the face plays a causal role in determining emotional experience. Because facial expression can be controlled voluntarily, some argue that there can be no systematic relationship with emotional experience. Supporters of the view that facial expressions of emotion evolved solely for communicative purposes believe that the link to emotional experience is epiphenomenal at best and perhaps even nonexistant (e.g., Mandler 1975). Others, invoking a cathartic-hydraulic model, claim that there is an inverse relationship between the outward expression of an emotion and the inward experience accompanying it, usually meaning autonomic reaction (Freud 1946; Jones 1950). From this point of view, the inhibition of facial expression intensifies, rather than attenuates, emotional reactions. Support for this hypothesis (e.g., Buck et al. 1974; Lanzetta and Kleck, 1970) led Buck (1980) to label what he called the "between-subjects" version vs the "within-subjects" version of the facial feedback hypothesis. The former pertains to the inverse relationship between expressivity and autonomic activity found across subjects, such as along the externalizer-internalizer distinction. The "within-subjects" version refers to those effects of facial expression on subjective experience across situations within individuals. The latter are the effects of interest here and are those most often associated with the facial feedback hypothesis.

This sampling of ideas from the past reveals a tradition of attributing a central role to somatic activity in the experience of an emotion and particularly to the patterned action of the facial musculature. Unfortunately, the theories differ widely in their specificity regarding both the primacy of the face and the mechanisms proposed to account for its effect, as well as in specifying the conditions under which the effect will or will not obtain. Consequently, laboratory research has arisen from a variety of hypotheses and has included diverse procedures both to manipulate expressive behavior and to evoke emotion. When Tourangeau and Ellsworth (1979) provided the field with three versions of the facial feedback hypothesis, and conducted an experiment that tested but failed to support any of them, a flurry of theoretical clarifications followed and subsequent research has been organized around more precise hypotheses.

Tourangeau and Ellsworth proposed that necessity, sufficiency, and monotonicity captured the relationships implied by existing theory and research. The necessity hypothesis states that the appropriate facial expression for an emotion is necessary for its subjective experience. The sufficiency hypothesis tests whether the voluntary production of a facial expression of emotion is sufficient to produce the subjective experience of that emotion. These versions of the hypothesis are most often attributed to Tomkins and Izard, who were very explicit about the role of the face in emotion, but who are often quoted out of context, making much of this research "strawmanish." These two versions assume that patterned activity of the facial muscles, even if voluntarily produced, leads directly to subjective experience of the appropriate emotion, and presumably to accompanying autonomic responses, even if there is no evocative stimulus. In the presence of a stimulus, the posed and the spontaneous expressions should blend in subjective experience, additively within an emotion, but in as yet unspecified ways for incongruent emotions.

The monotonicity hypothesis states simply that the relationship between facial expression and subjective experience is monotonic and positive. This reflects the Darwinian proposition that voluntary facilitation or inhibition of an evoked expressive response will change subjective experience, and presumably autonomic responses, in an additive way. Each hypothesis has unique implications for research strategy, and two primary paradigms have emerged in which they have been tested.

The first major line of research has explored how simulation of a facial expression influences self-report and autonomic reactions, thereby testing the sufficiency hypothesis: if subjects pose facial expressions and then are left alone or are presented with reliable elicitors of emotion, will their verbal report and autonomic reactions be modified? The second line of research concerns the effect of modulation of an evoked facial expression on the intensity of the emotional reaction to the eliciting stimulus, thereby testing the monotonicity hypothesis: If subjects exaggerate or inhibit their facial expression of an evoked emotion, in an attempt to either reveal or conceal it, then will that likewise affect their autonomic reactions and self-reports of their experience?

In the original experiments using the simulation paradigm, Laird (1974) instructed subjects to move specific facial muscles to produce partial smiles and frowns and found effects on both judgments of the stimuli and on mood ratings, following exposure to pleasant and unpleasant slides. Subjects who were "frowning" rated slides of KKK meetings more negatively and their own mood as more aggressive; subjects who were smiling rated slides of children playing more positively and rated their own mood as more pleasant. Follow-up research by Laird has generally supported the sufficiency hypothesis, but there have been conspicuous failures to support this version of the facial feedback hypothesis (e.g., McCaul et al. 1982; Tourangeau and Ellsworth 1979). Also, Winton (1986) has argued that Laird's effects reduce to a positive-negative dimensional level, and that evidence for simulation effects on self-reports of discrete emotional experience still do not exist.

In another test of the sufficiency hypothesis, Ekman et al. (1983) found discrete autonomic patterning associated with posed facial expressions. Precise muscle-by-muscle instructions were given to experienced subjects until they produced a prototypical expression of an emotion. Heart rate and skin temperature changes revealed three discrete reactions: a) low heart rate: happiness, disgust, surprise; b) high heart rate/low skin temperature: fear, sadness; c) high heart rate/high skin temperature: anger. To our knowledge, self-report data have not been collected with the directed facial action task, nor have subjects been posed and then exposed to an elicitor; thus, the links to Laird's research are missing.

In the first experiments with the modulation paradigm, Lanzetta et al. (1976) found that the exaggeration or suppression of a pain display in anticipation, or receipt, of electric shock, respectively, augmented or reduced skin conductance responses and reported painfulness, compared to the spontaneous expression of pain. When subjects were told to conceal their expression of pain so that someone watching a videotape of their reaction could not judge the shock intensity, they showed a reduced skin resistance change and reported less shock painfulness than they did when they did not control their expression. Likewise, when subjects were told to exaggerate their reactions to the shock so judges of their facial expressions would think it was very painful, their skin resistance changes were larger and their self-reports of shock painfulness were higher. Figure 11 shows the results for subjective reports and Fig. 12 for skin conductance.

Follow-up research by Lanzetta and Kleck (Colby et al. 1977; Kleck, et al 1976; Lanzetta et al. 1982; Vaughan and Lanzetta 1981) has supported the monotonicity hypothesis, both when instructions to hide or reveal facial expression have been used, as well as when context manipulations, such as the presence of an observer or of a mirror, have been used to inhibit expressive behavior. Hide/reveal instructions were used by Zuckerman et al. (1981) in a deception paradigm and by Kraut (1982), who used olfactory

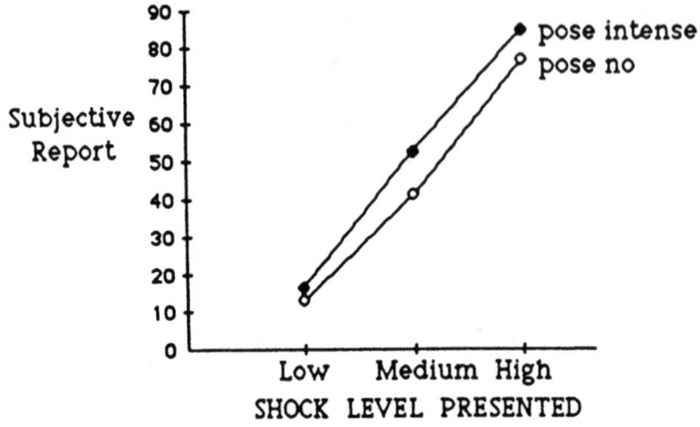

Fig. 11. Mean subjective reports of shock painfulness, Experiment 3. (Adapted from Lanzetta et al. 1976)

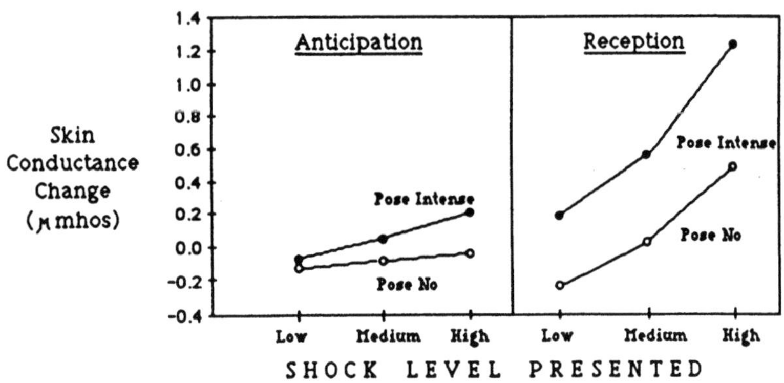

Fig. 12. Mean phasic conductance change scores to shock anticipation and shock reception, Experiment 3. (Adapted from Lanzetta et al. 1976)

stimuli to elicit affective reactions; these studies provided partial support for the effects of modulating facial expression on self-reported and autonomic reactions.

What are the mechanisms proposed to underlie the effects of facial feedback? That is, what is the nature of the information from the face, and how does it influence those events in the brain that create the experience of emotion? Tomkins (1962, 1963) and Gellhorn (1964), following Allport (1924), proposed that the peripheral mechanism involves patterned proprioceptive feedback from the muscles of facial expression. Because in this view each discrete emotion is associated with a unique facial expression, these theorists proposed that afferent information from the source of the expression, the facial muscles, is used centrally to differentiate emotional states, although their theories differ as to the brain centers and pathways involved. Tomkins

(1980) has recently revised his theory to include somatosensory afferents from the skin of the face, in part because the spindle density in the facial muscles is low and thus proprioceptive feedback to the brain may be too coarse for its imputed role.

In a similar vein, when Zajonc (1985) recently revived Waynbaum's turn-of-the-century theory concerning the relationship of facial expression to cerebral blood flow, he suggested that facial expressions that differentially affect cerebral blood flow in turn affect brain temperature, which may provide the mechanism for facial feedback effects. Finally, among those invoking a physiological mechanism, Ekman (see Ekman et al. 1983) believes that the feedback itself need not arise peripherally, that is, from the face, but rather the efferent commands from the motor cortex that lead to facial expression may be input to hypothalamic processes that differentiate one emotion from another. If so, emotions in dreams when muscule movement is inhibited, or emotions in clinical populations where facial movement or sensation is abnormal, are no longer problems for theories of facial feedback, but at the same time, the proposed mechanism becomes largely untestable.

In contrast to these physiologically based mechanisms, Laird (1974, 1984) has adopted a self-perception framework within which to view facial feedback, moving the mechanism into the cognitive realm. In effect, if you perceive yourself smiling, then you conclude that you feel happy; if you perceive yourself frowning, then you conclude that you must be angry. Lanzetta and Kleck (Kleck et al. 1976) proposed a conditioning model to explain their modulation effects: repeated exposure to emotional situations where both the rapid-acting facial expression and the slower-acting autonomic reaction are unconditioned responses may fortuitiously condition the autonomic reaction to the facial expression. Thus, apart from any innate tendencies, the dynamic properties of the two subsystems may lead to interoceptive conditioning, whereby expressive responses gain control over autonomic, and hence subjective, responses.

Such is the variety of mechanisms proposed to account for facial feedback effects. Unfortunately, there are no data that allow strong inferences concerning the underlying mechanisms at this time.

In summary then, our brief survey supports the recent reviews of the facial feedback research (e.g., Manstead 1988; Matsumoto1987; Rutledge and Hupka 1985; Whissell 1985) which have concluded that there is good support for the monotonicity hypothesis, weak support for the sufficiency hypothesis, and little or no support for the necessity hypothesis. However, meta-analysis quickly reveals that the empirical support is difficult to evaluate because the research has been plagued with ill-specified hypotheses and diverse methodologies.

There is a need for improved manipulations of facial behavior. Instructions to move select facial muscles, or to hide or reveal facial expressions, are subject to criticism on the grounds of experimental demand. That is, subjects may deduce the purpose of the experiment, especially using within-subject designs, and then modify behavior (usually verbal) accordingly. Yoked-control subjects and autonomic measures have been used to soften this criticism, but better manipulations may avoid it entirely. For example, Kleck et al. (1976) used a contextual manipulation, the presence of an observer, to induce subjects to inhibit their expressive behavior, while anticipating and receiving electric shock.

On the positive side, the recent emergence of facial EMG procedures promises to improve both real-time and post hoc verification of the intended facial behavior, thus providing important detail concerning the manipulation. In addition, new techniques for visually scoring facial muscle actions such as FACS (Ekman and Friesen 1978) can be used to verify facial manipulations from a videotape record of the experiment.

There is also need to expand upon our means for evoking emotion in the laboratory. For example, the initial experiments using the modulation paradigm involved electric shock and the dissimulation of the pain grimace. The electric shock insured an emotional response in subjects and allowed hypothesis testing over controlled levels of stimulus intensity, but pain is not considered a discrete emotion by many theorists, which makes the link to facial feedback hypotheses questionable to them. Therefore, modulation effects need to be tested using proximal stimuli that evoke direct, fundamental emotions, but few such attempts have been made other than Kraut's use of strong odors to evoke pleasure or disgust (Kraut 1982). We need procedures that are active rather than passive, that evoke emotions that are direct rather than vicarious, unless specific predictions for vicarious emotion are specified (e.g., Vaughan and Lanzetta 1981). Low intensity affect may be a fact of life in today's psychology laboratory, but there may be ways to make the emotional experience more natural.

Several additional issues have been raised, but not yet addressed, by research thus far, such as the voluntary vs involuntary source of facial expression. The conscious modulation of expressive behavior following instructions to do so may have different implications for emotional experience than modulation that is produced by unobtrusive means. Likewise, modulation of a spontaneously evoked facial expression may have different effects than modulation of a self-generated or simulated expression. Indeed, Leventhal (1984) has argued that the subjective experience of emotion arises from a discrepancy between intended expression and actual expression, that is, when involuntary, spontaneous expressive behavior collides with the "expected" expression. In the same way that some theorists (e.g., Hager and Ekman 1981) have insisted upon the accurate simulation of prototypical facial expressions of emotion for adequate tests of the facial feedback hypothesis, investigators should recognize the dual innervation of the facial muscles. That is, apart from being isomorphic with the prototype, was the facial expression produced spontaneously or volitionally? Another such case is that of stimulus forewarning, which is especially important for studies using the modulation paradigm. It is possible that knowledge about the nature and onset of an evocative stimulus may strongly influence attempts to control the emotion. Once an emotion is elicited and coping responses have begun, including autonomic changes, it may be less likely that modifying facial expression will effectively alter the natural course of the emotion.

Finally, we believe that the facial feedback hypothesis has been wrongly extracted from the larger context, where emotion is considered to result from the actions of several interacting subsystems. The expressive system is only one component in the complex process that determines emotional experience. Most theorists include the autonomic system, the somatic system, which includes expression and instrumental behavior, and subjective experience as minimal components in a theory of emotion. In a given situation, the coupling among the subsystems will vary, and the extent of this coupling may moderate any effects on emotional experience due to simulation or modulation of facial expression. In this larger context, both the static and the dynamic properties of a given system, such as the cardiovascular system or the hormonal system, determine what influence somatic activity, such as modulation of facial expression, can have in the overall process called emotion. Progress toward understanding the relationship between facial expression and emotional experience will be greater when facial feedback hypotheses are viewed as part of a larger network of propositions that interrelate the diverse components of the emotion process.

Acknowledgements. The research reported here was supported by the National Science Foundation, the Harry Frank Guggenheim Foundation, and the Lincoln Filene Endowment to Dartmouth College. The authors gratefully acknowledge the contributions of Lauren Bush to the preparation of this manuscript.

REFERENCES

Allport FH (1924) Social psychology. Houghton Mifflin, Boston

Ax AF (1953) The physiological differentiation between fear and anger in humans. Psychosom Med 15: 433-442

Buck R (1980) Nonverbal behavior and the theory of emotion: the facial feedback hypothesis. J Pers Soc Psychol 38: 811-824

Buck R, Savin V, Miller R, Caul W (1972) Communication of affect through facial expressions in humans. J Pers Soc Psychol 23: 362-371

Buck R, Miller R, Caul W (1974) Sex, personality, and physiological variables in the communication of affect via facial expression. J Pers Soc Psychol 30: 587-596

Cacioppo JT, Petty RE (1979) Attitudes and cognitive response: an electrophysiological approach. J Pers Soc Psychol 37: 2181-2199

Cacioppo JT, Petty RE, Marshall-Goodell B (1984) Electromyographic specificity during simple physical and attitudinal tasks: location and topographical features of integrated EMG responses. Biol Psychol 18: 85-121

Cacioppo JT, Petty RE, Losch ME, Kim H (1986) Electromyographic activity over facial muscle regions can differentiate the valence and intensity of affective reactions. J Pers Soc Psychol 50: 260-268

Chance MRA (1976) Social attention: society and mentality. In: Chance MRA, Larsen RR (eds) The social structure of attention. Wiley, New York, p 315

Colby CZ, Lanzetta JT, Kleck RE (1977) Effects of the expression of pain on autonomic and pain tolerance responses to subject-controlled pain. Psychophysiology 14: 537-540

Darwin C (1965) The expression of the emotions in man and animals. University of Chicago Press, Chicago (Originally published 1872)

Ekman P (1972) Universals and cultural differences in facial expressions of emotion. In: Cole J (ed) Nebraska symposium on motivation, 1971 (vol 9). University of Nebraska Press, Lincoln

Ekman P, Friesen W (1978) The facial action coding system. Consulting Psychologists, Palo Alto, Ca

Ekman P, Oster H (1979) Facial expressions of emotion. Annu Rev of Psychol 30: 527-554

Ekman P, Sorenson E, Friesen W (1969) Pan-cultural elements in facial displays of emotion. Science 164: 86-88

Ekman P, Friesen W, Tomkins S (1971) Facial affect scoring technique (FAST): a first validity study. Semiotica 3: 37-38

Ekman P, Friesen W, Ellsworth P (1972) Emotion in the human face: guidelines for research and an integration of the findings. Pergamon, Elmsford New York

Ekman P, Friesen W, Ancoli S (1980) Facial signs of emotional experience. J Pers Soc Psychol 39: 1125-1134

Ekman P, Levenson R, Friesen W (1983) Autonomic nervous system activity distinguishes among emotions. Science 221: 1208-1210

Englis B, Vaughan K, Lanzetta JT (1982) Conditioning of counter-empathetic emotional responses. J Exp Soc Psychol 18: 375-391

Freud S (1946) Instincts and their vicissitudes. In: Collected papers, vol 4. Hogarth, London (Originally published 1921)

Fridlund A, Schwartz G, Fowler S (1984) Pattern recognition of self-reported emotional state from multiple-site facial EMG activity during affective imagery. Psychophysiology 21: 622-636

Gellhorn E (1964) Motion and emotion: the role of proprioception in the physiology and pathology of the emotions. Psychol Rev 71: 457-472

Hager JC, Ekman P (1981) Methodological problems in Tourangeau and Ellsworth's study of facial expression and experience of emotion. J Pers Soc Psychol 40: 358-362

Izard CE (1971) The face of emotion. Appleton-Century-Crofts, New York

Izard CE (1979) The maximally discriminative facial movement coding system (MAX). University of Delaware, Newark

James W (1962) Psychology: briefer course. Collier, New York

Jones HE (1950) The study of patterns of emotional expression. In: Reymert M (ed) Feelings and emotions. McGraw-Hill, New York

Kleck R E, Vaughan R C, Cartwright-Smith J, Vaughan, K B, Colby C Z, Lanzetta

J T (1976) Effects of being observed on expressive, subjective, and physiological responses to painful stimuli. J Pers Soc Psychol 34: 1211-1218
Kraut RE (1982) Social presence facial feedback and emotion. J of Pers Soc Psychol 42: 853-863
Laird JD (1974) Self-attribution of emotion: the effects of expressive behavior on the quality of emotional experience. J Pers Soc Psychol 29: 475-486
Laird JD (1984) The real role of facial response in the experience of emotion: a reply to Tourangeau and Ellsworth and others. J Pers Soc Psychol 47: 909-917
Lanzetta JT (1976) Effects of being observed on expressive subjective and physiological responses to painful stimuli. J Pers Soc Psychol 34: 1211-1218
Lanzetta JT, Kleck RE (1970) Encoding and decoding of nonverbal affect in humans. J Pers Soc Psychol 16: 12-19
Lanzetta JT, Cartwright-Smith J, Kleck RE (1976) Effects of nonverbal dissimulation on emotional experience and autonomic arousal. J Pers Soc Psychol 33: 354-370
Lanzetta JT, Biernat JJ, Kleck RE (1982) Self-focused attention, facial behavior, autonomic arousal, and the experience of emotion. Motivation Emotion 6: 49-63
Lanzetta JT, Sullilvan DG, Masters RD, McHugo GJ (1985) Emotional and cognitive responses to televised images of political leaders. In: Krauss S, Perloff RM (eds) Mass media and political thought: toward an information processing approach. Sage, Beverly Hills, p 85
Leventhal H (1984) A perceptual motor theory of emotion. In: Scherer KR, Ekman P (eds) Approaches to emotion. Lawrence Erlbaum, Hillsdale, p 271
Leventhal H, Brown D, Shacham S, Engquist G (1979) Effect of preparatory information about sensations threat of pain and attention on cold pressor distress. J Pers Soc Psychol 37: 688-714
Mandler G (1975) Mind and emotion. Wiley, New York
Manstead ASR (1988) The role of facial movement in emotion. In: Wagner HL (ed) Social psychophysiology: theory and clinical applications. Wiley, Chichester, p 105
Matsumoto D (1987) The role of facial response in the experience of emotion: more methodological problems and a meta-analysis. J Pers Soc Psychol 52: 769-774
McCaul KD, Holmes DS, Solomon S (1982) Voluntary expressive changes and emotion. J Pers Soc Psychol 42: 145-152
McHugo GJ (1983) Facial EMG and self-reported emotion. Paper presented at the 23rd annual meeting of the Society for Psychophysiological Research, Asilomar, CA
McHugo G, Smith C, Lanzetta J (1982) The structure of self-reports of emotional responses to film segments. Motivation Emotion 6: 365-385
McHugo GJ, Lanzetta JT, Sullivan DG, Masters RD, Englis BG (1985) Emotional reactions to a political leader's expressive displays. J Pers Soc Psychol 49:1512-1523

McHugo GJ, Lanzetta JT, Bush LK (1988) The effect of attitudes on emotional reactions to expressive displays of political leaders. (to be published)

Ohman A, Fredrikson M, Hugdahl K, Rimmo P (1976) The premise of equipotentiality in human classical conditioning: conditioned electrodermal responses to potentially phobic stimuli. J Exp Psychol Gen 105: 313-337

Redican WK (1982) An evolutionary perspective on human facial displays. In Ekman P (ed) Emotion in the human face, 2nd eds. Cambridge University Press, Cambridge, p 212

Rutledge LL, Hupka RB (1985) The facial feedback hypothesis: methodological concerns and new supporting evidence. Motivation Emotion 9: 219-240

Schachter S, Singer J (1962) Cognitive, social, and physiological determinants of emotional state. Psychol Rev 69: 379-399

Schwartz GE, Fair PL, Salt P, Mandel MK, Klerman GL (1976) Facial muscle patterning to affective imagery in depressed and non-depressed subjects. Science 192: 489-491

Smith C, McHugo G, Lanzetta J (1986) The facial muscle patterning of posed and imagery-induced expressions of emotion by expressive and non-expressive posers. Motivation Emotion 10: 133-157

Tomkins SS (1962) Affect, imagery, consciousness, vol 1. The positive affects. Springer, New York

Tomkins SS (1963) Affect, imagery, consciousness, vol 2. The negative affects. Springer, New York

Tomkins SS (1980) Affect as amplification: some modifications in theory. In Plutchik R, Kellerman H (eds) Emotion: Theory, research, and experience, vol 1, Theories of emotion. Academic Press, New York, p 141

Tourangeau R, Ellsworth P (1979) The role of facial response in the experience of emotion. J Pers Soc Psychol 37:1519-1531

Van Hooff JARAM (1969) The facial displays of the Catarrhine monkeys and apes. In: Morris D (ed) Primate ethology. Anchor, New York, p 9

Van Hooff JARAM (1973) A structural analysis of the social behavior of a semi-captive group of chimpanzees. In: von Cranach M, Vine I (eds) Social communication and movement. Academic Press, New York, p 75

Vaughan KB, Lanzetta JT (1980) Vicarious instigation and conditioning of facial expressive and autonomic responses to a model's expressive display of pain. J Pers Soc Psychol 38: 909-923

Vaughan KB, Lanzetta JT (1981) The effect of modification of expressive displays on vicarious emotional arousal. J Exp Soc Psychol 17: 16-30

Whissell C (1985) The role of the face in human emotion: first system or one of many? Percept Mot Skills 61: 3-12

Winton W (1986) The role of facial response in self-reports of emotion: a critique of Laird. J Pers Soc Psychol 50: 808-812

Zajonc RB (1985) Emotion and facial efference: a theory reclaimed. Science 228: 15-21

Zuckerman M, Klorman R, Larrance D, Spiegel N (1981) Facial autonomic and subjective components of emotion: the facial feedback hypothesis versus the externalizer/internalizer distinction. J Pers Soc Psychol 41: 929-944

Components of Emotional Behavior Disrupted in Patients with Unilateral Brain Lesions

What can be Learned from Lateral Differences in Emotional Processing?

P. Feyereisen
University of Louvain, Belgium

INTRODUCTION

Over the last few years, a considerable body of data has been gathered showing differences in the roles of the two cerebral hemispheres in emotional processing. Studies on lateral differences in normal subjects and in subjects with affective illnesses have been added to the previous observations on the behaviour of brain-damaged people (see the other chapters of this volume and the reviews by Bruyer 1980; Campbell 1982a; Davidson 1984; Gainotti 1983, 1984; Heilman et al. 1983; Kinsbourne and Bemporad 1984; Seron and Vanderlinden 1979; Silberman and Weingartner 1986; Tucker 1981; Wexler 1980). In the field of emotions as elsewhere (e.g. Hellige 1983), interactions between tasks and hemispheres demonstrate that related processes may dissociate, so there is a way of delineating psychological functioning. Two questions may be examined from such a perspective (Feyereisen 1988). Are emotional processes distinct from other modes of information processing? What relations exist between the different emotional processes involved in the production and comprehension of verbal, vocal, facial and physiological signals?

The Specificity of Emotional Processing: Emotion and Cognition

Specifying the lateral differences in emotional processing can lead to several interpretations. If one considers that these differences concern only emotions, the lateralisation of emotions would not relate to hemispheric specialisation for language, motor control, or visual processes. Right-hemisphere dominance in emotional processing would not depend on its superiority in visuospatial processing or on lateral biases in attention. Another conception attributes asymmetries in emotional processes to the more general characteristics of the two hemispheres that are revealed in several cognitive processes. For instance, Tucker (1981) interprets lateral differences in emotional processing as another illustration of the specialisation of the left and right hemispheres in, respectively, analytic and holistic modes of processing. Other hypotheses relate lateral differences in emotional processing to interhemispheric asymmetries in attention and control (Tucker and Williamson 1984; Kinsbourne and Bemporad 1984).

Experimental Brain Research Series 18

The interpretation of lateral differences in emotional processing is thus related to the issue of the separability of emotion and cognition. Is emotional information processing special and does it parallel cognitive processing? Or does this distinction correspond to other subdivisions in the cognitive domain? These questions will first be examined from theoretical perspectives by analysing different conceptions of emotions and seeing what is attributed to each side of the brain. Second, empirical data on the emotional and cognitive processing of facial, vocal and verbal information by normal and brain-damaged subjects will be reviewed.

Theories of Emotions and Lateral Differences

The study of lateral differences in emotional processing invites us to draw connections between several fields of investigation. On the one hand, the lateralisation of emotion is related to other investigations of emotion. In this regard, concern in recent years has been mainly directed to criticism of the theory proposed by Schachter and Singer (1962): according to them emotional experience results from a cognitive interpretation of physiological arousal in its context of occurrence. Many hypotheses derived from the cognition-arousal theory were not supported by empirical data, so alternative views were proposed (Cotton 1981; Leventhal and Tomarken 1986; Manstead and Wagner 1981; Parkinson 1985; Reisenzein 1983; Zillman 1983). The analysis of lateral differences in emotional processing ought to be affected by this evolution.

On the other hand, there are links between this analysis and discussions on lateral differences in perceptual and motor processes. In this area, the respective competences of the two hemispheres are no longer considered as relating to the verbal or non-verbal nature of the stimulus in a unitary way. Instead, they are thought to depend on task characteristics and on psychophysical input parameters (Sergent and Hellige 1986; Hellige and Sergent 1986). In the visual processing of words or faces, lateral differences vary with regard to the processing demands. From such a perspective, a general right-hemisphere dominance for emotional processing is not expected, but it should be modified according to experimental conditions.

In approaching the problem of lateralisation of emotion, we must first define the processes under analysis and identify the roles played in them by the two hemispheres. The different "theories" of emotion that psychology has proposed all assume that two modes of information processing, emotion and cognition, can be distinguished. They diverge in the way these processes are characterised.

Emotions as Bodily Responses

Sensory input may activate different kinds of representations in parallel: verbal labels or conceptual nodes within a associative network on the one hand, and diverse autonomic and motor responses on the other. The bodily manifestations do not depend on a deep cognitive interpretation of the eliciting conditions. Zajonc and Markus (1984) call the bodily responses that signal the occurrence of emotional information "hard" representations of affect as opposed to the "soft" representations constituted by subjective experience. Other authors make different distinctions (Leventhal 1984; Buck 1985). Emotion then consists of different kinds of processes that generate the production of vegetative reactions, expressive behaviour and "syncretic" subjective experience. These emotional responses would compete with conceptual processes that control voluntary motor activity and verbal reports of emotion. Conceptual representations, indeed, are involved in the application of "display rules" when emotions are feigned or repressed. The emotional mode of processing is concerned with the automatic activation of motor programmes and of "emotional schemata" coding subjective experience in analogic forms; the conceptual mode deals with the knowledge about emotions that can be expressed in words or controlled action.

From such perspectives, lateral differences may be used to distinguish modes of processing. However, the distinction between left- and right-hemisphere functioning may be delineated in different ways. Zajonc (1984) opposes emotional and cognitive processes without giving precise definitions of the terms: "emotional reactions are likely to be under the control of the right brain hemisphere, whereas cognitive processes are predominatly the business of the left hemisphere" (p.119). A less crude distinction is drawn by Leventhal (1984) and Buck (1985), for whom subjective experience is a particular kind of cognitive process. Thus, following Tucker (1981), the division of labour they see between the two hemispheres opposes two modes of cognition: "the right hemisphere acts as a synthetiser, organizing input into complex and global wholes, and this kind of syncretic conceptualization is involved in emotion" (Buck 1985, p. 395). Similarly, according to Leventhal (1984), lateral differences concern the distinction between two memory systems, a concrete one involving imagery and an abstract one built from propositional representations. Thus, it is suggested that lateral differences in emotional processing relate to the analytic/holistic dichotomy and to the hemispheric specialisation in mental imagery. These hypotheses deserve discussion.

The distinction between analytic and holistic modes of processing was introduced to account for the effects of task demands on lateral differences. One or the other hemisphere may be advantaged by processing the same verbal or non-verbal stimulus according to the requested response or the way the subject performs the task. This manner of characterising hemispheric specialisation has been challenged (Bradshaw and Nettleton 1981). Indeed, it is considered vain to search for a single label to cover all the processes for which one hemisphere shows an advantage (Bertelson 1982). The circularity of the reasoning has also been criticised: in the absence of external criteria for an a priori distinction between analytic and holistic processes, one could call all the processes for which a right-hemisphere advantage is observed holistic (Sergent 1983). An alternative would be to consider that the hemispheres show different sensitivity to the physical characteristics of the input. More particularly, the right hemisphere would be less impaired than the left in processing degraded information such as a visual presentation with a short duration of exposure, in the periphery of the visual field, at a low luminance level, or after brain damage (Sergent 1984). Lateral differences in emotional processing have not yet been analysed from such a perspective. Another answer to the problem of circularity is to define holistic processes as bearing on structural or configurational features and analytic processes on local and partial features (Morais 1982). This approach has not been used in the study of emotional processes.

The analysis of lateralisation of mental imagery processes does not support the hypothesis of a right-hemisphere dominance without qualification (Ehrlichman and Barrett 1983; Farah 1984). Imagery involves several components, some of which rely on categorical representations and the semantic network activated during verbal production. These imagery processes would give rise to a left-hemisphere advantage (Kosslyn 1987). Accordingly, some reliance on mental imagery during emotional processing would not account unambiguously for right-hemisphere superiority.

Cognitive Theories of Emotions

Cognitive and somatic theories of emotion agree on the distinctions between various modes of information processing. The main differences bear on the description of the processes by which some conditions elicit emotional responses: somatic theories are not explicit about the way in which bodily reactions are triggered, whereas cognitive theories suppose that overt emotional behaviour results from a preliminary analysis of the situation. Several hypotheses have been proposed stressing different aspects of the response production mechanism. For some authors, emotional reactions follow successive appraisals of the eliciting conditions in several dimensions such as novelty, pleasantness and goal relevance (Frijda 1987; Leventhal and Scherer 1987; Scherer 1984; Smith and Ellsworth 1985). Others consider emotion as an interruption procedure that allows organisms to pursue different goals in a partially unpredictable environment: happiness, fear and anger are ways of inhibiting continuing action and organising priorities among competing plans (Oatley and Johnson-Laird 1987; Mandler 1980; Simon 1967). These rapid emotional processes differ from more durable affective states like anxiety or depression which also interact with cognitive processes in state-dependent memory tasks (Bower and Cohen 1982; Blaney 1986). Subjects in "feeling of knowing" states are also able to retrieve emotional attributes of items they can no longer recall (Schacter and Worling 1985). Such phenomena may be explained by assuming a nonzero rest level of activation of associative nodes in a semantic network. From this point of view, emotion results from placing input in correspondence with recollected autobiographical events.

Within these cognitive approaches of emotion, little attention has been given to possible hemispheric specialisation. Nevertheless, these cognitive conceptions are relevant for a neuropsychology of emotion. The fractionation of emotional reactions into elementary components permits a more precise description of the respective role of the two hemispheres. If the emotional meaning of a stimulus depends on an evaluation of familiarity, pleasantness and other dimensions, lateral differences could result from hemispheric specialisation for these early processes rather than from specialisation for the emotional response. For example, some data from outside the field of emotion suggest a different influence of familiarity on information processing by the right and the left hemisphere (see e.g. Bruyer 1986). If emotion stems from a judgment of novelty and if the right hemisphere shows a superiority in processing unfamiliar events, it could be more involved than the left in emotional processing. Similarly, interactions between laterality effects and the positive or negative valence of the stimulus suggest different capacities or biases of the two hemispheres in the appraisal of pleasantness (Davidson 1984). Furthermore, if multiple memory stores are assumed, it might happen that affectively loaded events are represented in the right hemisphere. For instance, recollection of sad feelings slows down reaction times to stimuli presented to the left visual field (right hemisphere) more than to the right (Ladavas et al. 1984). The hypothesis of right-hemisphere involvement in emotional storage also accounts for lateral differences in the influence of mood on the recognition of emotional facial expressions (Gage and Safer 1985).

Lateral differences in emotional processing might also depend on a more general specialisation if emotion represents an emergency solution when planned action is interrupted. Another way of characterising the hemispheric functioning underlines the dynamic aspects of information processing (for reviews, see Cohen 1982; Allen 1983). From such a perspective, lateral differences appear when attention paid to the continuing processes varies across the hemispheres. In fact, the notion of attention relates to several components: alerting the organism, selecting the information and allocating resources (Posner 1975; Pribram and McGuiness 1975). Tucker and Williamson (1984) have suggested, on the basis of neurochemical and behavioural data, that the right hemisphere was more readily aroused than the left, but less able to control attention. This hypothesis, however, remains speculative and studies on lateral differences in attention do not support such a general claim without qualification. Allocation of a resource to one or the other hemisphere might depend on the nature of the information processed (Kinsbourne and Hiscock 1983; Moscovitch 1986). Attentional deficits caused by right parietal lesions seem to result from impaired selective attention rather than from a reduced state of alertness; arousal would depend on subcortical structures rather than on the right parietal cortex (Posner et al. 1984; Butter 1987; Riddoch and Humphreys 1987).

The Comprehension of Emotional Cues

The analysis of the processes by which emotion is perceived in other people suggest that different levels of comprehension may be involved. A primitive mode of understanding could rely on an automatic response to emotional signals in the same way as releasing mechanisms described in ethology trigger preprogrammed reactions. For instance, exposure to a facial expression elicits spontaneous mimicry (Dimberg 1982; Vaughan and Lanzetta 1980). Laughter and crying are also contagious in some circumstances (e.g. Leventhal and Cupcik 1975). These empathic reactions demonstrate a elementary kind of emotional information processing.

However, in some conditions the attribution of emotional states to other people results from an inferential process. Therefore, other mechanisms than those giving rise to emotional reactions are involved. Assuming, on the basis of situational and behavioural cues, pain, sadness or pleasure in the mind of somebody else does not imply that similar feelings must be experienced at the same time by oneself. Instead, these assumptions probably rely on an implicit "theory of mind" that, like other kinds of knowledge, develops during childhood (see e.g. Bretherton, Fritz et al. 1986; Gnepp and Gould 1985; Harris 1983; Strayer 1986).

The distinction between a "direct" perception of emotional cues and more deductive procedures is part of a larger theoretical debate (Fodor and Pylyshyn 1981; McArthur and Baron 1983; Turvey et al. 1981; Ullman 1980; Wagemans 1986). The problem is to explain how a sensory pattern appearing in a noisy environment and with continuously varying components gives rise to a discrete, categorical response. Words, faces and objects are recognised as invariants in spite of differences in appearance and in contexts of occurrence. Only a few attempts have been made to analyse from such a perspective the processes by which vocal and facial expressions of emotions are categorised (e.g. Russel and Fehr 1987). Nevertheless, the interpretation of lateral differences depends on the model proposed to describe these operations.

Emotional and Cognitive Processing of Faces and Speech

The question of the specificity of emotion has inspired comparisons of laterality effects in emotional and non-emotional processes. These studies bear principally on facial, vocal and verbal behaviour. Psychophysiological correlates of cognitive processing do not seem to have attracted similar attention, in spite of the interest that the analysis of these aspects in brain-damaged people may offer.

Facial Asymmetries and Lateral Differences in Face Perception

Some facial movements express emotions, while others relate various cognitive processes like perplexity or skepticism. Similarly, the face provides emotional and non-emotional information about such things as identity, gender and age. Thus, the analysis of facial processing allows one to examine the question of whether lateral differences are influenced by the emotional character of the process.

One body of evidence concerns the production of facial movements by normal subjects. The asymmetry of the face expressing an emotion, which appears to be more intense on the left side that is under direct right-hemisphere control, than on the right side, is now well-documented but still a matter of discussion (for reviews, see Borod and Koff 1984; Campbell 1986; Hager 1982; Sackeim and Gur 1983; Thomson 1985). Among the still unsolved problems is whether the asymmetry is related to emotion or, more generally, to motor control. In some cases, the expected interaction between laterality effect and the emotional nature of the movement has been observed (Cacioppo and Petty 1981). In others, the interaction concerned the emotional valence of the expression (e.g. Sackeim and Grega 1987). It does not seem that differences in expressivity rely on the anatomical asymmetry of the face at rest (Sackeim 1985). However, they might depend on a greater mobility of the left side of the face (Koff et al. 1981). Similar asymmetries were observed in the production of non-emotional movements (Alford and Alford 1981; Alford 1983; Borod and Koff 1983; Campbell 1982b). Inversely, according to Ekman et al. (1981), genuine emotional expressions are symmetrical, and only poses show lateral differences. This hypothesis, however, was not supported by the observation of spontaneous asymmetrical expressions (Borod et al. 1983; Dopson et al. 1984; Tucker et al. 1985). Furthermore, many facial movements produced on request were symmetrical (Hager and Ekman 1985). Lateral differences also appeared in the kind of movement produced, but the factors favouring the right or the left side of the face remain unknown. The study of mouth asymmetry raises other interpretation problems (Graves 1983; Graves et al. 1982; Hager and Van Gelder 1985; Graves et al. 1985). Greater mouth opening during speech may either reflect higher activation of the contralateral hemisphere during verbal production or it may be related to the left-hemisphere dominance in the control of speech and non-speech motor sequences (Kimura 1982; Wolf and Goodale 1987).

Difficulties in interpreting facial asymmetries arise partially from our ignorance about the functional value of differences in expressivity. Does a more intense movement of one side of the face depend on a greater activation of the contralateral hemisphere or on a greater inhibition of the opposite side (Rinn 1984)? Does it show an "advantage" or less precise control? To some extent, observations of brain-damaged subjects are less problematic because unilateral lesions necessarily impair facial behaviour and because the severity of the deficit after right and left lesions can be compared. Unfortunately, studies contrasting the production of emotional and non-emotional facial movements are rare and no conclusions can as yet be drawn (Feyereisen 1986). Furthermore, the data on group comparisons in producing facial movements on request are contradictory. However, posing for emotions seems to dissociate from performing orofacial movements in apraxia examinations (Borod et al. 1986; Pizzamiglio et al. 1987). Moreover, in left-hemisphere-damaged subjects, facial movements were performed more correctly when the command stressed their emotional meaning than when neutral terms were used (Borod et al. 1987). In this experiment, right-hemisphere-damaged subjects showed no impairments under both emotional and neutral conditions. Facial asymmetry studies in brain-damaged subjects also indicate that the two hemispheres intervene in facial motor control (Bruyer 1981). Mouth opening in aphasic subjects was influenced by the nature of the task: smiling, singing and counting up to ten (automatic speech) were left sided while naming, answering questions and repetition were right sided (Graves and Landis 1985). The process behind these asymmetries, hemispheric activation by task or constraints in motor control, remains unknown.

The question of the specificity of emotional facial processes has been more clearly answered on the receptive level. Many arguments indicate that recognising facial identity, is dissociated from interpreting emotional expressions. For example, stimulus familiarity influenced facial identity judgment but not emotional identity judgment in a study using the central presentation of faces to normal subjects (Young et al. 1986). Covariance analysis on laterality effects in perception of expression and identity (e.g. Ley and Bryden 1979; Pizzamiglio et al. 1983) and studies on brain-damaged subjects (e.g. Etcoff 1984a; Bowers and Heilman 1984; Bowers et al. 1985) have also showed the independence of the two processes. These data are reviewed in more detail elsewhere and will not be discussed further here (Feyereisen 1986; Etcoff, this volume).

Emotional and Linguistic Processing of Vocal and Verbal Information

Like the face, speech can, by its prosodic and segmental features, transmit emotional and non-emotional information. The quality of voice expresses emotions, underlines the semantic focus of an utterance and gives indications of the identity of the speaker. Similarly, some words refer to emotional state or to affectively loaded matters, while others express neutral meanings. To some extent, music presents the same characteristics: a given piece may elicit emotion as well as "colder" judgments.

Dichotic listening studies in normal subjects do not provide a clear indication that lateral differences depend on the emotional nature of the task or the input. The data showing a right-hemisphere advantage in emotional processing are difficult to interpret (Bryden et al. 1982; Ley and Bryden 1982). Indeed, similar left-ear (right-hemisphere) superiority appeared under different conditions, not only in the perception of emotional vocal signals like laughter and crying, but also in recognising the noises of coughing or sneezing, in identifying the speaker's age and gender and in processing the intonation contour of sentences (e.g. Blumstein and Cooper 1974; King and Kimura 1972; Carmon and Nachshon 1973). Furthermore, the left-ear advantage in the recognition of emotional sounds partially depends on attentional strategies (Mahoney and Sainsbury 1987). A task in which subjects were requested to rate the pleasantness of musical passages did not change lateral differences in comparison with a control condition (Peretz et al. 1987).

Both hemispheres seem to be involved in the emotional processing of verbal stimuli, whereas the linguistic processing of the same material gives rise to left-hemisphere dominance. The left-ear superiority observed in reacting to the prosody of neutral words disappeared when words with emotional meaning were used (Morais and Ladavas 1987; Zenhausern et al. 1981) and when the emotional content of sentences presented monaurally was rated (Safer and Leventhal 1977). However, in a spoken-word recognition task, the usual right-ear superiority was not influenced by the emotional character of the words (Morais and Ladavas 1987). In lexical decisions in the visual modality, the results of Graves et al. (1981) showing a lower right-field advantage for emotional than for neutral words in male subjects were not replicated (Strauss 1983). In female subjects, emotional load related to performance in the right but not the left visual field, and the recognition of words presented in the left visual field was influenced by their imagery value (Deloche et al. 1987; Graves et al. 1981). In brain-damaged patients, similarity judgments bearing on written words did not show a clear dissociation in the processing of the connotative and the denotative aspects of meaning (Brownell et al. 1984). Patients suffering from right-hemisphere lesions were mainly influenced by antonymy relations (e.g. "warm-cold"); in this respect, they behaved like normal control subjects and furthermore, it cannot be asserted that they did not rely on the emotional meaning of these words. In a memory task, both right- and left-hemisphere damaged patients recalled better neutral texts than emotional ones (Wechsler 1973).

The behaviour of patients suffering from unilateral brain damage does not show a clearer dissociation of emotional and linguistic processing of speech. Right-hemisphere-damaged subject are not only impaired in processing emotional prosody but also in perceiving linguistic stress (Weintraub et al. 1981), in recognising speaker voices (Assal et al. 1981; Van Lancker and Canter 1982), and in understanding or producing the interrogative raise (Heilman et al. 1984; Shapiro and Danly 1985). Furthermore, the comprehension of emotional prosody is also impaired in left-hemisphere-damaged subjects, often in association with auditory comprehension deficits (Schlanger et al. 1976; Tompkins and Flowers 1985). Some data show that processing semantic content interact with processing prosody in identifying the emotional meaning of an utterance. Aphasics, even in cases of severe comprehension disorders, and right-hemisphere- damaged subjects processed prosody better when the semantic content was congruent rather than in conflict with the tone of voice (Bowers et al. 1987; Seron et al. 1982). These results do not support a model that assumes both an absolute hemispheric specialisation in emotional and linguistic processing and a total suppression of these processes in cases of brain lesions. Two interpretations remain possible: either the two hemispheres are involved in the processing of emotional sentences or the damaged hemisphere is still able to process partially the information for which it is specialised. The dichotic listening procedure allows a comparison of the two hemispheres in brain-damaged subjects. In a recognition task of emotional and linguistic prosody, right- and left-hemisphere-damaged subjects did not differ from normal subjects when the performance of the intact hemisphere was analysed; however, the right-hemisphere- damaged subjects did show a slight, insignificant impairment. Both groups were impaired in the processing of stimuli presented to the ear contralateral to the lesion (Hartje et al. 1985). Thus, it was again shown that both hemispheres are involved in the recognition of prosody.

The last field in which the question of the specificity of the right hemisphere in emotional processing arises is the understanding of humour in jokes or cartoons (Gardner et al. 1975). The analysis of errors made by right-hemisphere- damaged subjects indicates that the impairment concerns only the ability to find a coherent link between the beginning of the story and the end (Bihrle et al. 1986; Brownell et al. 1983). In verifying the relations between sentences and emotional context, these subjects also tended to accept incongruous conclusions (Tompkins and Mateer 1985). A similar impairment emerged when the task was simply to find a logical but not funny ending of a story (Bihrle et al. 1986). Right-hemisphere-damaged subjects were sensitive to the comic character of a situation when perceptual incongruities were presented (Dagge and Hartje 1985; Bihrle et al. 1986). Thus, the inability to understand humorous stories might result from a more general deficit in the inability to draw inferences (Gardner et al. 1983; Brownell et al. 1986). The very nature of this deficit, however, is as yet unclear. Right-hemisphere-damaged subjects acknowledged the truth of correct inferences but failed to reject incorrect ones (McDonald and Wales 1986; Joanette and Goulet, personal communication). Thus, they made inferences, albeit along sometimes bizarre lines.

Conclusions

The question of specificity in emotional processing cannot be unequivocally answered if it is cast in general terms. A review of the literature leads to different conclusions, depending on the field of inquiry. It has not yet been demonstrated that facial asymmetry or lateral differences in processing prosody are related to the emotional nature of the information, but it is very likely that comprehension of emotional facial expression relies on specific mechanisms. Experimental testing of hypotheses on the influence of task and input factors on emotional processes also requires different control conditions with regard to the proposed material. Thus, describing the right hemisphere as specialised in the holistic mode of processing does not seem any more warranted than suggesting a specific involvement with emotion. In both cases, a detailed examination of the task demands is required. In order to benefit fully from observations on lateral differences, a more precise description of the operations underlying the production and comprehension of emotional signals is needed.

The Diversity of Emotional Processes

Emotion manifests itself in physiological reactions, vocal and facial expressions, and verbal reports describing a subjective experience. Theories of emotion often assume that these emotional cues represent different facets of a single phenomenon. Thus, hemispheric specialisation for emotion per se should involve all these components to the same extent. Nevertheless, lateral differences may affect the diverse emotional responses in distinct ways. First, these emotional reactions may rely on different cerebral bases. They may also vary along a control dimension. It is often assumed that the emotional quality of a behaviour result from automatic releases: emotions differ from controlled actions by their obligatory character. However, in many experimental tasks, the subjects are asked to pose for emotions or to match intonation and picture, without necessarily experiencing the emotion. Thus, dissociations may occur between automatic and controlled processing of emotional information or between on-line and off-line measurements.

Automaticity and Control

Although the clinical evidence for a dissociation between automatic and voluntary activity has been acknowledged in the neuropsychological literature ever since Jackson (1961), few systematic descriptions of these phenomena are available. The situation is slowly changing in the linguistic and perceptual domains, but little support for the hypothesis has been provided in the experimental analysis of emotions. For instance, the laterality effect does not seem to be influenced by the spontaneous or posed nature of facial expressions of emotions (e.g. Borod et al. 1983, 1986; Dopson et al. 1984; Pizzamiglio and Mammucari, this volume). Other kinds of dissociation have been demonstrated, however, notably by Caltagirone and Zoccolotti (this volume), who showed the independence between physiological responses and spontaneous facial expression during a film presentation.

Few attempts have been made to compare automatic and controlled processes in the perception of emotional cues: it remains to be proved that right-hemisphere-damaged people who are impaired in matching facial and vocal expressions of emotions do not react to these signals in daily life situations. Cases exist in which physiological measures show reactions to stimuli that cannot be consciously identified (e.g. Bauer 1984, 1986; Tranel and Damasio 1985; Tranel et al. 1985). Similarly, amnesics sometimes show behavioural reactions to stimuli previously associated with emotional information, in spite of an inability to retrieve that information in a memory task (Johnson et al. 1985).

Contemporary descriptions of automatic processes rely on the assumption of autonomous modules specialised in input-output transfer functions. Moreover, it is suggested that some direct "routes" connect perceptual recognition devices to response systems without activating cognitive representations or requiring attentional resource. Thus, the description of automatic emotional reactions depends on the identification of these autonomous components. Neuropsychology may contribute here by testing provisional models that assume several levels of information processing in the production and comprehension of emotional signals.

Lateral Differences and Levels of Emotional Processing

The aim of studies on the lateralisation of psychological functioning is to analyse a given behaviour in terms of processes taking place between signal occurrence and response production. It is assumed that hemispheric lesions and lateral differences in normal subjects do not affect the different components of the process in the same way. This approach is taken by

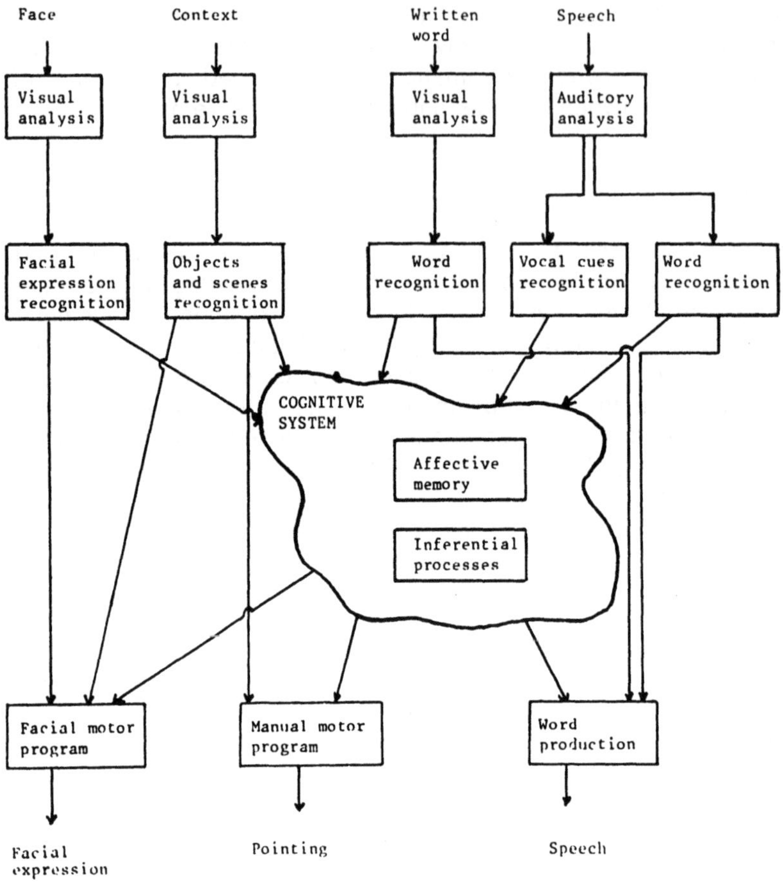

Fig.1 A simplified model of the operations involved in the comprehension of emotional signal by brain-damaged subjects, based on the Logogen model (Warren and Morton, 1982)

numerous studies on reading, face and object recognition, and speech production but has found little application in the neuropsychology of emotion. Nevertheless, a tentative model may be proposed by analogy with other schematic descriptions of picture naming and word reading.

Fractionation of Facial and Verbal Emotional Processing

Information processing models first assume that a signal undergoes modality specific perceptual analysis (Fig. 1). For the sake of simplicity, only visual and auditory signals are considered here, but tactile, olfactory and gustatory signals that convey emotional information could obviously also be represented by such a conception. Just as words activate lexical nodes, hypothetical processes might involve special recognition devices for facial and vocal expressions of emotion. The mechanism by which these graded signals are identified as belonging to discrete categories, as well as the more general process of categorising natural objects, remains disputed (Smith and Medin 1981). A classical approach suggests that these signals are recognised on the basis of their constituent features: they are analysed as a set of more elementary units. Alternatively, classification may result from a judgment of similitude with prototypes. In this case, the incoming stimulus would activate templates of basic and typical exemplars of joy, anger, fear, etc. Another possibility would be an evaluation of the stimulus in function of affective dimensions that are not modality specific such as the pleasantness dimension, on which all kinds of stimuli may be rated (e.g. Green and Cliff 1975). Verbal and non-verbal signals also contact, within the cognitive system, modality-specific representations like autobiographic memories and propositional representations, allowing for intermodal integration and for inference from situational cues. Finally, diverse components control the production of verbal, motor or autonomic responses. It may be assumed that motor and autonomic responses do not require a mediation of the cognitive system in some conditions, but result directly from a categorization or a dimensional appraisal of the stimulus. Similarly, pleasantness rating, like other non-emotional judgments, may be made before full recognition of the signal (e.g. Mandler et al. 1987).

The schematic distinctions between several components of emotional information processing (Fig.1) are intended only to offer a more precise description of the operations involved in the use of facial, vocal and verbal signals and to specify the nature of impairments observed in brain-damaged subjects. They should first inspire studies comparing comprehension and production of emotional signals. Several authors have stressed the interest, equalling that of lateralisation studies, of differentiations along the anteroposterior or dorsoventral axis of neuroanatomical structures involved in emotional processing (Davidson 1984; Kinsbourne and Bemporad 1984; Bear 1983). Of course, we are far from being able to specify the way in which the different components of emotions are implemented in the human brain (LeDoux, 1986). These hypotheses, however, deserve attention as they suggest that the perceptual analysis, motor control and cognitive interpretation implied by emotions rely on different mental functions. Several case studies have shown dissociations between input and output systems in processing affective prosody (Ross 1981; Gorelick and Ross 1987; Speedie et al. 1984). Similarly, in the group of brain-damaged subjects sampled by Borod et al. (1986), performances in facial-expression comprehension and production did not correlate.

Furthermore, some patients may suffer from modality-specific impairments. Right-sided lesions may disturb emotional processing of faces or other visual stimuli but spare the understanding of verbal expressions of emotion (Habib 1986; Kolb and Taylor 1981). The patient described by Bowers and Heilman (1984) could not match facial expressions and names of emotions, but he could process emotional intonation correctly. Similar cases, or cases showing the inverse dissociation, were reported in the series of Benowitz et al.

(1983). As far as output systems are concerned, right-hemisphere-damaged subjects viewing slides and describing emotional experience were impaired in their vocal and facial but not verbal expression (Borod et al. 1985).

Comprehension of Emotional Signals in Brain-Damaged Subjects: a Pilot Study

These considerations suggest that the comprehension of facial expression of emotion may be impaired in different ways after right- or left-hemisphere lesions. A pilot study was conducted to examine two hypotheses; one assuming a modality-specific deficit in categorising facial expressions and the other a more general supramodal deficit affecting emotional processing (Feyereisen P. Meurens P. and Mijnsberghe K., unpublished data, 1987).

Three experiments were conducted with eight right- hemisphere-damaged and eight left-hemisphere-damaged subjects. The subjects were not selected for particular symptoms, but they were given a detailed neuropsychological examination. A pretest analysed the ability to draw inferences in right-brain-damaged subjects (Brownell et al. 1986; McDonald and Wales 1986). The left-brain damaged subjects were given other pretests on reading and auditory comprehension. The small size of the samples did not create the most important problem in interpreting the results. The tested hypotheses are not incompatible, and it may well be that one subject suffered from a specific disorder while another showed supramodal deficits or an association of several impairments. The heterogeneity of clinical populations has methodological consequences on testing hypotheses in group studies (Caramazza 1986). The statistical comparison of two samples of unilaterally brain-damaged patients gives little useful information. It only allows us to estimate the proportion of impaired subjects of different kinds in the right- and left-brain-damaged groups; however, even a small number of pathological cases after a left hemisphere lesion would suffice to show an involvement of that hemisphere in the behaviour under analysis. To obtain indications on the organisation of the processes underlying the comprehension of facial expression, individual performances must be examined.

The three experiments followed similar procedures. The subjects were asked to select from a vertical array of three items (presented on a page on the right) the response that corresponded to the target on the left. In the first experiment, the comprehension of facial expressions was compared with that of words referring to emotional state. These conditions should differ in cases of modality-specific impairments but not if a supramodal deficit affects emotional processing. The first block of 24 target faces expressing joy, sadness or anger was presented along with the response faces of different actors. It was followed by a second block of 24 written target words referring to the same emotional states. More errors occurred in the face than in the word part, both in the right-hemisphere-damaged subjects (Wilcoxon $T = 0$, $p = 0.01$) and in the six cases of left-hemisphere-damaged subjects who completed the task (Wilcoxon $T = 0$, $p = 0.06$). Errorless performances of right-brain-damaged subjects in the verbal part as compared with defective processing of facial expressions confirmed other observations showing that the semantic lexicon of the emotional terms was intact in these subjects (Etcoff 1984b; Kolb and Taylor 1981). Two interpretations are possible: either they suffered from a modality-specific impairment for faces or the task demands for processing words were reduced, as shown by the performances of the left-brain-damaged subjects. However, in normal subjects making semantic decisions about the emotional valence of faces from the Ekman series and words expressing joy or sadness, faster response times for faces than for words were observed (Feyereisen et al. 1986).

In the second experiment (Table 1), categorisation of facial expression was compared with a more inferential task relying on the cognitive system. The face part of the previous experiment was compared with a scene part: the 24 targets were pictures depicting situations in which joy, sadness, or anger is likely to be experienced by one character. As before, the responses had to be selected from faces. In neither group did the conditions differ significantly (Wilcoxon T = 12.5 and 11.0). Thus, it was not more difficult to match different facial expressions than to associate faces with situational contexts.

The third experiment was more particularly devoted to the analysis of the precategorisation process (Table 1). It was hypothesised that subjects with a specific disorder in the recognition of the facial expression of emotions remain able to process physical identity, whereas non-specific visual impairments disturb performances even when the decision does not require emotional processing. Bowers et al. (1985) found right- but not left-hemisphere-damaged subjects impaired concerning same-different judgments of emotional expression posed by the same actors. The two groups were impaired when judgments involved the expressions of different actors. The present study tried to determine whether the difficulty is in extracting invariant information from faces of different persons or just in detecting the physical identity of two exemplars of the same photographs. A control condition used different views of the same actor posing for the same emotion. Thus, three blocks of 16 stimuli were compared. In each block, eight happy faces were randomly mixed with eight angry faces. In the first part (emotion identity), the correct response was the face expressing the same emotion as the target face posed by a different actor. In the second part (person identity), the correct response was a different view of the same target actor posing for the same emotion. In the last part (physical identity), the correct response was a copy of the target face. In each case, the distractors were two faces expressing another emotion, one by the target actor, the other by a different person. These three parts differed significantly in the left-hemisphere-damaged subjects only. In this group, more errors occurred in the emotion identity part than in the person identity and the physical identity parts (Friedman statistics = 7.00, p = 0.05). A similar analysis in the right-hemisphere-damaged group did not show significant differences between parts (Friedman statistics = 3.56). Subjects showing impairments in the physical identity part, in which the correct response may be given without emotional processing, might well suffer from an elementary perceptual deficit. However, errors more often consisted in choosing the same actor as

Table 1. Individual results in the face matching tasks

(total number of errors)

Subject	Experiment 2 (max = 24)		Experiment 3 (max = 16)		
	Faces	Scenes	Physical	Person identity	Emotion
Right brain-damaged subjects					
Subgroup A: no hemineglect, few errors in inference task					
CN	10	6	0	4	1
HT	12	7	0	0	0
CO	12	16	3	6	8
Subgroup B: no hemineglect, errors in inference task					
TT	7	7	6	6	7
Subgroup C: hemineglect, few errors in inference task					
GI	11	6	0	1	1
DZ	3	8	0	1	2
CT	12	14	8	4	2
Subgroup D: hemineglect, errors in inference task					
SO	13	11	0	0	5
Left brain-damaged subjects					
Subgroup A: Mild comprehension deficit, normal reading					
DZ	8	7	0	0	3
MX	5	6	0	0	O
WS	3	6	0	0	2
BS	4	8	0	0	0
Subgroup B: Moderate comprehension deficit, dyslexia					
FR	5	11	1	2	7
Subgroup C: Severe comprehension deficit, alexia					
CZ	9	7	2	3	4
DR	15	18	6	7	12
RN	15	7	7	2	9

the target with a different expression; thus, the misleading similarity in appearance between the target and the distractor might also have influenced the response selection rather than just the perceptual analysis of the input.

For no part of the three experiments did the differences between the right- and left-brain-damaged subjects reach significance (Mann-Whitney U tests), nor did emotional valence influence the results. The absence of differences between groups reflected the heterogeneity of the samples. However, examination of the individual performances did not demonstrate obvious dissociations between tasks (Table 1). It did not allow us to correlate the experimental results with the preliminary neuropsychological investigations. Impairments of facial processing were observed in almost all the right-brain-damaged subjects, with or without hemineglect or difficulties in drawing inferences. Similarly, in left-brain-damaged subjects, the performances did not relate to the extent of comprehension disorders. A severe global aphasic like CZ did not make more errors than DZ (subcortical damage) or FR, who suffered from a moderate Wernicke's aphasia.

It is also confirmed that left-hemisphere lesions impair the processing of emotional facial expressions. These subjects have been shown to perform worse than normal subjects but better than right-brain-damaged subjects (Cicone et al. 1980; DeKosky et al. 1980; Goldblum 1980; Kulikov and Sidorova 1983; Walker-Batson et al. 1987). It is usually concluded that the right hemisphere plays a dominant role in emotional processing. The facts, however, argue for renewed attention to the role of the left hemisphere. It has been suggested that a part of the problems encountered by aphasic subjects is in the inferential linkage between the nonverbal signal and the situational context (Feyereisen et al. 1981). However, in the second experiment, matching faces to emotional scenes did not appear more difficult than face-to- -face matching, and thus other factors relating to the categorisation process probably also intervene in the impairment displayed by aphasic subjects.

These results should not rule out further studies of single cases of impaired facial processing. The conditions of experiments 2 and 3 addressed the facial expression recognition stage. Moreover, the subjects were unselected, so the conditions for finding dissociations were not met. A preferable research strategy might probably be to analyse in depth the conditions under which impairments in facial processing occur. This conclusion also applies to the study of emotion comprehension in other clinical populations, for example, schizophrenia, dementia or mental handicap. Indeed, it is very unlikely, that all the pathological cases in which the interpretation of emotional facial expressions is disturbed suffer from a deficit of the same nature.

Conclusions

The initial observations on lateral differences in emotional processing have opened the way to new questions: is the hemispheric specialisation domain-specific or does it result from more general properties of the cerebral organisation? No unambiguous solution to this problem can be provided without subdivisions within emotional behaviour. Input and output modalities have to be distinguished to analyse the specificity issue in a more detailed way. But then other problems arise: how are the different components related? Are the different processes lateralised, i.e. impaired by unilateral brain lesions, to the same extent? From such a perspective, the study of lateral differences shows that the notion of "emotion" refers to a great variety of phenomena that experimental neuropsychology can help to distinguish.

REFERENCES

Alford R D (1983) Sex differences in lateral facial facility: the effects of habitual facial concealment. Neuropsychologia 21: 567-570

Alford R , Alford K F (1981) Sex differences in asymmetry in the facial expressions of emotion. Neuropsychologia 19: 605-608

Allen M (1983) Models of hemispheric specialization. Psychol Bull 93: 73-104

Assal G , Aubert C , Buttet J (1981) Asymétrie cérébrale et reconnaissance do la voix. Rev Neurol 137: 255-268

Bauer R M (1984) Autonomic recognition of names and faces in prosopagnosia: a neuropsychological application of the Guilty Knowledge Test. Neuropsychologia 22: 457-469

Bauer R M (1986) The cognitive psychophysiology of prosopagnosia. In: Ellis H D , Jeeves M A , Newcombe F Young A (eds) Aspects of face processing. Nijhoff, Dordrecht, pp 253-267

Bear D M (1983) Hemispheric specialization and the neurology of emotion. Arch Neurol 40: 195-202

Benowitz L I, Bear D M, Rosenthal R, Mesulam M M, Zaidel E, Sperry R W (1983) Hemispheric specialization in nonverbal communication. Cortex 19: 5-11

Bertelson P (1982) Lateral differences in normal man and lateralization of brain function. Int J Psychol 17: 173-210

Bihrle A M, Brownell H H, Powelson J A, Gardner H (1986) Comprehension of humorous and nonhumorous materials by left and right brain-damaged patients. Brain Cogn 5: 399-411

Blaney P H (1986) Affect and memory: a review. Psychol Bull 99: 229-246

Blumstein S, Cooper W E (1974) Hemispheric processing of intonation contour. Cortex 10: 146-158

Borod J C, Koff E (1983) Hemiface mobility and facial expression asymmetry. Cortex 19: 327-332

Borod J C, Koff E (1984) Asymmetries in affective facial expression: behavior and anatomy. In: Fox N, Davidson R (eds) The psychobiology of affective development. Erlbaum, Hillsdale, 293-323

Borod J C, Koff E, White B (1983) Facial asymmetry in posed and sponatenous expressions of emotions. Brain Cogn 2: 165-175

Borod J C, Koff E, Perlman-Lorch M, Nicholas M (1985) Channels of emotional expressions in patients with unilateral brain damage. Arch Neurol 42: 345-348

Borod J C, Koff E, Perlman-Lorch M, Nicholas, M (1986) The expression and perception of facial emotion in brain-damaged patients. Neuropsychologia 24: 169-180

Borod J C, Perlman-Lorch M, Koff E, Nicholas M (1987) Effect of emotional context on bucco-facial apraxia. J Clin Exp Neuropsychol 9: 155-161

Bower G H, Cohen P R (1982) Emotional influences in memory and thinking: data and theory. In: Clark M S, Fiske S T (eds) Affect and cognition. Erlbaum, Hillsdale, pp 291-331

Bowers D, Heilman K M (1984) Dissociation between the processing of affective and nonaffective faces: a case study. J Clin Neuropsychol 6: 367-379

Bowers D, Bauer R M, Coslett H B, Heilman K M (1985) Processing of faces by patients with unilateral hemisphere lesions. I. Dissociation between judgments of facial affect and facial identity. Brain Cogn 4: 258-272

Bowers D, Coslett H B, Bauer R M, Speedie L J, Heilman K M (1987) Comprehension of emotional prosody following unilateral hemispheric lesions: processing defect versus distraction defect. Neuropsychologia 25: 317-328

Bradshaw J L, Nettleton N C (1981) The nature of hemispheric specialization in man. Behav Brain Sci 4: 51-91

Bretherton I, Fritz J, Zahn-Waxler C, Ridgeway D (1986) Learning to talk about emotions: a functionalist perspective. Child Dev 57: 529-548

Brownell H H, Michel D, Powelson J, Gardner H (1983) Surprise but not coherence: sensitivity to verbal humor in right-hemisphere patients. Brain Lang 18: 20-27

Brownell H H, Potter H H, Michelow D, Gardner H (1984) Sensitivity to lexical denotation and connotation in brain-damaged patients: a double dissociation? Brain Lang 22: 253-265

Brownell H H, Potter H H, Bihrle A M, Gardner, H (1986) Inference deficits in right brain-damaged patients. Brain Lang 27: 310-321

Bruyer R (1980) Implication différentielle des hémisph}res cérébraux dans les conduites émotionnelles. Acta Psychiatr Belg 80: 266-284

Bruyer R (1981) Asymmetry of facial expression in brain damaged subjects. Neuropsychologia 19: 615-624

Bruyer R (1986) L'hémisph}re cérébral gauche est-il chez l'homme davantage dépendant du temps que le droit? Année Psychol 86: 247-259

Bryden M P, Ley R G, Sugarman J H (1982) A left-ear advantage for identifying the emotional quality of tonal sequences. Neuropsychologia 20: 83-87

Buck R (1985) Prime theory: an integrated view of motivation and emotion. Psychol Rev 92: 349-413

Butter C M (1987) Varieties of attention and disturbances of attention: a neuropsychological analysis. In: Jeannerod M (ed) Neurophysiological and neuropsychological aspects of spatial neglect. North Holland, Amsterdam, pp 1-23

Cacioppo J T, Petty R E (1981) Lateral asymmetry in the expression of cognition and emotion. J Exp Psychol Hum Percept 7: 333-341

Campbell R (1982a) The lateralisation of emotion: a critical review. Int J Psychol 17: 211-219

Campbell R (1982b) Asymmetries in moving faces. Br J Psychol 73: 95-103

Campbell R (1986) Asymmetries of facial action: Some facts and fancies of normal face movement. In: Bruyer R (ed) The neuropsychology of face perception and facial expression. Erlbaum, Hillsdale, pp 247-267

Caramazza A (1986) On drawing inferences about the structure of normal cognitive systems from the analysis of patterns of impaired performances: the case for single patients studies. Brain Cogn 5: 41-66

Carmon A, Nachshon I (1973) Ear asymmetry in perception of emotional nonverbal stimuli. Acta Psychol 37: 351-357

Cicone M, Wapner W, Gardner H (1980) Sensitivity to emotional expressions and situations in organic patients. Cortex 16: 145-158

Cohen G (1982) Theoretical interpretations of lateral asymmetries. In: Beaumont J G (ed) Divided visual field studies of cerebral organisation. Academic, London, pp 87-111

Cotton J L (1981) A review of research on Schachter's theory of emotion and the misattribution of arousal. Eur J soc Psychol 11: 365-397

Dagge M, Hartje W (1985) Influence of contextual complexity on the processing of cartoons by patients with unilateral lesions. Cortex 21: 607-616

Davidson R J (1984) Affect, cognition, and hemispheric specialization. In: Izard C E, Kagan J, Zajonc R B (eds) Emotion, cognition, and behavior. Cambridge University Press, Cambridge, pp 320-365

DeKosky S T, Heilman K M, Bowers D, Valenstein E (1980) Recognition and discrimination of emotional faces and pictures. Brain Lang 9: 206-214

Deloche G, Seron X, Scius G, Segui J (1987) Right hemisphere language processing: lateral difference with imageable and nonimageable ambiguous words. Brain Lang 30: 197-2O5

Dimberg U (1982) Facial reactions to facial expressions. Psychophysiology 19: 643-647

Dopson W G, Beckwith B E, Tucker D M, Bullard-Bates P C (1984) Asymmetry of facial expression in spontaneous emotion. Cortex 20: 243-251

Ehrlichman H, Barrett J (1983) Right hemisphere specialization for mental imagery: a review of the evidence. Brain Cogn 2: 55-76

Ekman P, Hager J, Friesen W V (1981) The symmetry of emotional and deliberate facial actions. Psychophysiology 18: 101-106

Etcoff N L (1984a) Selective attention to facial identity and facial emotion. Neuropsychologia 22: 281-295

Etcoff N L (1984b) Perceptual and conceptual organization of facial emotions: hemispheric differences. Brain Cogn 3: 385-412

Farah M J (1984) The neurological basis of mental imagery: a componential analysis. Cognition 18: 245-272

Feyereisen P (1986) Production and comprehension of emotional facial expressions in brain-damaged patients. In: Bruyer R (ed) The neuropsychology of face perception and facial expression. Erlbaum, Hillsdale, pp 221-245

Feyereisen P (1988) Theories of emotions and neuropsychological research. In: Gainotti G (ed) Emotional behavior and its disorders. Elsevier, Amsterdam, (Handbook of neuropsychology, Vol VI) (in press)

Feyereisen P, Seron X, de Macar M A (1981) L'interprétation de différentes

catégories de gestes chez des sujets aphasiques. Neuropsychologia 19: 515-521
Feyereisen P, Malet C, Martin Y (1986) Is the faster processing of expressions of happiness modality-specific? In: Ellis H D, Jeeves M A, Newcombe F, Young A W (eds) Aspects of face processing. Martinus Nijhoff, Dordrecht, pp 349-355
Fodor J A, Pylyshyn Z W (1981) How direct is visual perception? Some reflections on Gibson's "ecological approach" Cognition 9: 139-196
Frijda N H (1987) Emotion, cognitive structure, and action tendency. Cognition and Emotion, 1: 115-143
Gage D F, Safer M A (1985) Hemisphere differences in the mood state-dependent effect for recognition of emotional faces. J Exp Psychol Learn Mem Cogn 11: 752-763
Gainotti G (1983) Laterality of affect: the emotional behavior of right- and left-brain-damaged patients. In: Myslobodsky M S (ed) Hemisyndromes: psychobiology, neurology, psychiatry. Academic, New York, pp 175-192
Gainotti G (1984) Some methodological problems in the study of the relationships between emotion and cerebral dominance. J Clin Neuropsychol 6: 111-121
Gardner H, Ling P K, Flamm L, Silverman J (1975) Comprehension and appreciation of humorous material following brain damage. Brain 98: 399-412.
Gardner H, Brownell H H, Wapner W, Michelow D (1983) Missing the point: The role of the right hemisphere in the processing of complex linguistic material. In: Perecman E (ed) Cognitive processing in the right hemisphere. Academic, New York, pp 169-191
Gnepp J, Gould M E (1985) The development of personalized inferences: understanding other people's emotional reactions in light of their prior experiences. Child Dev 56: 1455-1464
Goldblum M C (1980) La reconnaissance des expressions faciales émotionnelles et conventionnelles au cours des lésions cérébrales. Revue Neurol 136: 711-719
Gorelick P B, Ross E D (1987) The aprosodias: further functional-anatomical evidence for the organisation of affective language in the right hemisphere. J Neurol Neurosurg Psychiatry, 50: 553-560
Graves R (1983) Mouth asymmetry, dichotic ear advantage and tachistoscopic visual field advantage as measures of language lateralization. Neuropsychologia 21: 641-649
Graves R, Landis T (1985) Hemispheric control of speech expression in aphasia: a mouth asymmetry study. Arch Neurol 42: 249-251
Graves R, Landis T, Goodglass H (1981) Laterality and sex differences for visual recognition of emotional and nonemotional words. Neuropsychologia 19: 95-102
Graves R, Landis T (1985) Hemispheric control of speech expression in aphasia: a mouth asymmetry study. Arch Neurol 42: 249-251
Graves R, Landis C, Simpson C (1985) On the interpretation of mouth asymmetry. Neuropsychologia 23: 121-122
Green R S, Cliff N (1975) Multidimensional comparisons of stuctures of vocally and facially expressed emotion. Percept Psychophys 17: 429-438

Habib M (1986) Visual hypoemotionality and prosopagnosia associated with right temporal lobe isolation. Neuropsychologia 24: 577-582
Hager J C (1982) Asymmetries in facial expression. In: Ekman P (ed) Emotion in the human face, 2nd edn. Cambridge University Press, Cambridge, pp 318-352
Hager J C, Ekman P (1985) The asymmetry of facial actions is inconsistent with models of hemispheric specialization. Psychophysiology 22: 307-318
Hager J, Van Gelder R (1985) Asymmetry of speech action. Neuropsychologia 23: 119-120
Harris P H (1983) Children's understanding of the link between situation and emotion. J Exp Child Psychol 36: 490-509
Hartje W, Willmes K, Weniger D (1985) Is there parallel and independent hemispheric processing of intonational and phonetic components of dichotic speech stimuli? Brain Lang 24: 83-99
Heilman K M, Watson R T, Bowers D (1983) Affective disorders associated with hemispheric disease In: Heilman K M, Satz P (eds) Neuropsychology of human emotion. Guilford, New York, pp 45-64
Heilman K M, Bowers D, Speedie L, Coslett H B (1984) Comprehension of affective and nonaffective prosody. Neurology 34: 917-921
Hellige J B (1983) Hemisphere x task interaction and the study of laterality. In: Hellige J B (ed) Cerebral hemisphere asymmetry: method, theory, and application. Praeger, New York, pp 411-443.
Hellige J B, Sergent J (1986) Role of task factors in visual field asymmetries. Brain Cogn 5: 200-222
Jackson J H (1961) On affections of speech from disease of the brain. In: Shipley T (ed) Classics in psychology. Philosophical Library, New York, pp 582-612 (First published in 1879).
Johnson M K, Kim J K, Risse G (1985) Do alcoholic Korsakoff's syndrome patients acquire affective reactions? J Exp Psychol Learn Mem Cogn 11: 22-36
Kimura D (1982) Left-hemisphere control of oral and brachial movements and their relation to communication. Philos Trans R S Lond (Biol) 298: 135-149
King F L, Kimura D (1972) Left-ear superiority in dichotic perception of vocal nonverbal sounds. Can J Psychol 26: 111-116
Kinsbourne M, Bemporad B (1984) Lateralization of emotion: a model and the evidence. In: Fox N A, Davidson R J (eds) The psychobiology of affective development. Erlbaum, Hillsdale, 259-291
Kinsbourne M, Hiscock M (1983) Asymmetries of dual-task performances. In: Hellige J B (ed) Cerebral hemisphere asymmetry: method, theory, and application. Praeger, New York, pp 255-334
Koff E, Borod J C, White B (1981) Asymmetries for hemiface size and mobility. Neuropsychologia 19: 825-830
Kolb B, Taylor L (1981) Affective behavior in patients with localized cortical excisions: role of lesion side and site. Science 214: 89-91
Kosslyn S M (1987) Seeing and imaging in the cerebral hemispheres: a computational approach. Psychol Rev 94: 148-175
Kulikov M A, Sidorova O A (1985) Identification of intonational and facial expressions of emotion by patients with organic lesions of the right or the left hemisphere. Soviet Neurology and Psychiatry 18: 75-87

Ladavas E, Nicholetti R, Umilta C, Rizzolatti G (1984) Right hemisphere interference during negative affect: a reaction time study. Neuropsychologia 22: 479-485

LeDoux J E (1986) The neurobiology of emotion. In LeDoux J E, Hirst W (eds) Mind and brain: dialogues in cognitive neuroscience. Cambridge University Press, Cambridge, pp 3O1-354.

Leventhal H (1984) A perceptual-motor theory of emotion. In: Berkowitz L (ed) Advances in experimental social psychology, vol 17. Academic, New York, pp 117-182

Leventhal H, Cupchik G (1975) The informational and facilitative effects of an audience upon expression and the evaluation of humorous stimuli. J Exp Soc Psychol 11: 363-380

Leventhal H, Scherer K (1987) The relationship of emotion to cognition: a functional approach to a semantic controversy. Cognition and Emotion 1: 3-28

Leventhal H, Tomarken A J (1986) Emotion: today's problems. Annu Rev Psychol 37: 565-610

Ley R G, Bryden M P (1979) Hemispheric differences in processing emotion and faces. Brain Lang 7: 127-138

Ley R G, Bryden M P (1982) A dissociation of right and left hemispheric effects for recognizing emotional tone and verbal content. Brain Cogn 1: 3-9

Mahoney A M, Sainsbury R S (1987) Hemispheric asymmetry in the perception of emotional sounds. Brain Cogn 6: 216-233

Mandler G (1980) The generation of emotion: a psychological theory. In: Plutchik R, Kellerman H (eds) Theories of emotion. Academic, New York, pp 219-243 (Emotion: theory, research, experience, vol 1)

Mandler G, Nakamura Y, Shebo Van Zandt BJ (1987) Nonspecific effects of exposure on stimuli that cannot be recognized. J Exp Psychol Learn Mem Cogn 13: 646-648

Manstead A S R, Wagner H L (1981) Arousal, cognition and emotion: an appraisal of two-factor theory. Curr Psychol Rev 1: 35-54

McArthur L Z, Baron R M (1983) Toward an ecological theory of social perception. Psychol Rev 90: 215-238

McDonald S, Wales R (1986) An investigation of the ability to process inferences in language following right hemisphere brain damage. Brain Lang 29: 68-8O

Morais J (1982) The two sides of cognition. In: Mehler J, Garrett M, Walkes E (eds) Perspectives in mental representations. Erlbaum, Hillsdale, pp 277-309

Morais J, Ladavas E (1987) Hemispheric interactions in the recognition of words and emotional intonations. Cognition and Emotion 1: 89-100

Morrow L, Vrtunski B, Kim Y, Boller F (1981) Arousal responses to emotional stimuli and laterality of lesion. Neuropsychologia 19: 65-71

Moscovitch M (1986) Afferent and efferent models of visual perceptual asymmetries: theoretical and empirical implications. Neuropsychologia 24: 91-114

Oatley K, Johnson-Laird P N (1987) Towards a cognitive theory of emotions. Cognition and Emotion 1: 29-50

Parkinson B (1985) Emotional effects of false autonomic feedback. Psychol Bull 98: 471-494
Peretz I, Morais J, Bertelson P (1987) Shifting ear differences in melody recognition through strategy inducement. Brain Cogn 6: 202-215
Pizzamiglio L, Zoccolotti P, Mammucari A, Cesaroni R (1983) The independence of face identity and facial expression recognition mechanisms: relationship to sex and cognitive style. Brain Cogn 2: 176-188
Pizzamiglio L, Caltagirone C, Mammucari A, Ekman P, Friesen W V (1987) Imitation of facial movements in brain damaged patients. Cortex 23: 207-221
Posner M I (1975) Psychobiology of attention. In: Gazzaniga M S, Blakemoore C (eds) Handbook of psychobiology. Academic, New York, pp 441-480
Posner M I, Walker J A, Friedrich F J, Rafal R D (1984) Effects of parietal injury on covert orienting of attention. J Neurosci 4: 1863-1874
Pribram K H, McGuiness D (1975) Arousal, activation, and effort in the control of attention. Psychol Rev 82: 116-149
Reisenzein R (1983) The Schachter theory of emotion: two decades later. Psychol Bull 94: 239-264
Riddoch M J, Humphreys G W (1987) Perceptual and action systems in unilateral visual neglect. In: Jeannerod M (ed) Neurophysiological and neuropsychological aspects of spatial neglect. North Holland, Amsterdam, pp 151-181
Rinn W E (1984) The neuropsychology of facial expression: a review of the neurological and psychological mechanisms for producing facial expressions. Psychol Bull 95: 52-77
Ross E D (1981) The aprosodias: functional-anatomic organization of the affective components of language in the right hemisphere. Arch Neurol 38: 561-569
Russell J A, Fehr B (1987) Relativity in the perception of emotion in facial expressions. J Exp Psychol (Gen) 116: 223-237
Sackeim H A (1985) Morphologic asymmetries of the face: a review. Brain Cogn 4: 296-312
Sackeim H A, Grega D M (1987) Perceiver bias in the processing of deliberately asymmetric emotional expression. Brain Cogn 6: 464-473
Sackeim H A, Gur R C (1983) Facial asymmetry and the communication of emotion. In: Cacioppo J T, Petty R E (eds) Social psychophysiology. Guilford, New York, pp 307-352
Safer M A, Leventhal H (1977) Ear differences in evaluating emotional tones of voice and verbal content. J Exp Psychol Hum Perception 3: 75-82
Schachter S, Singer J (1962) Cognitive, social, and physiological determinants of emotional state. Psychol Rev 69: 379-399
Schacter D L, Worling J R (1985) Attribute information and the feeling-of-knowing. Can J Psychol 39: 467-475
Scherer K R (1984) On the nature and function of emotion: a component process approach. In: Scherer K R, Ekman P (eds) Approaches to emotion. Erlbaum, Hillsdale, pp 293-317
Schlanger B B, Schlanger P, Gerstman LJ (1976) The perception of

emotionally toned sentences by right-hemisphere-damaged and aphasic subjects. Brain Lang 3: 396-403
Sergent J (1983) The analytic/holistic dichotomy: an epiphenomenon. Behav Brain Sci 6: 521-523
Sergent J (1984) Inferences from unilateral brain damage about normal hemispheric functions in visual pattern recognition. Psychol Bull 96: 99-115
Sergent J, Hellige J B (1986) Role of input factors in visual-field asymmetries. Brain Cogn 5: 174-199
Seron X, Vanderlinden M (1979) Vers une neuropsychologie humaine des conduites émotionnelles? Année Psychol 79: 229-252
Seron X, Van der Kaa M, Vanderlinden M, Remits A, Feyereisen P (1982) Decoding paralinguistic signals: effect of semantic and prosodic cues on aphasics' comprehension. J Commun Disord 15: 223-231
Shapiro B E, Danly M (1985) The role of the right hemisphere in the control of speech prosody in propositional and affective contexts. Brain Lang 25: 19-36
Silberman E K, Weingartner H (1986) Hemispheric lateralization of functions related to emotion. Brain Cogn 5: 322-353
Simon H A (1967) Motivational and emotional controls of cognition. Psychol Rev 74: 29-39
Smith C A, Ellsworth P C (1985) Patterns of cognitive appraisal in emotion. J Pers Soc Psychol 48: 813-838
Smith E E, Medin D L (1981) Categories and concepts. Harvard University Press, Cambridge
Speedie L J, Coslett H B, Heilman K M (1984) Repetition of affective prosody in mixed transcortical aphasia. Arch Neurol 41: 268-270
Strauss E (1983) Perception of emotional words. Neuropsychologia 21: 99-103
Strayer J (1986) Children's attributions regarding the situational determinants of emotion in self and in others. Dev Psychol 22: 649-654
Thomson J K (1985) Right brain, left brain; left face, right face: hemisphericity and the expression of facial emotion. Cortex 21: 281-289
Tompkins C A, Flowers C R (1985) Perception of emotional intonation by brain-damaged adults: the influence of task processing levels. J Speech Hear Res 20: 527-530
Tompkins C A, Mateer C A (1985) Right hemisphere appreciation of prosodic and linguistic indications of implicit attitude. Brain Lang 24: 185-203
Tranel D, Damasio A R (1985) Knowledge without awareness: An autonomic index of facial recognition by prosopagnosics. Science 228: 1453-1454
Tranel D, Fowles D C, Damasio A R (1985) Electrodermal discrimination of familiar and unfamiliar faces: a methodology. Psychophysiology 22: 403-408
Tucker D M (1981) Lateral brain function, emotion, and conceptualization. Psychol Bull 89: 19-46
Tucker D M, Williamson P A (1984) Asymmetric neural control systems in human self-regulation. Psychol Rev 91: 185-215
Tucker D M, Beckwith B E, Dopson W G, Bullard-Bates P C (1985)

Asymmetrical facial expression indeed: a reply to Dave and Thomson. Cortex 21: 305-307
Turvey M T, Shaw R E, Reed E S, Mace W M (1981) Ecological laws of perceiving and acting: in reply to Fodor and Pylyhyn (1981). Cognition 9: 237-304
Ullman S (1980) Against direct perception. Behav Brain Sci 3: 373-415
Van Lancker D R, Canter G J (1982) Impairment of voice and face recognition in patients with hemispheric damage. Brain Cogn 1: 185-195
Vaughan K B, Lanzetta JT (1980) Vicarious instigation and conditioning of facial expressive and autonomic response to a model's expressive display of pain. J Pers Soc Psychol 38: 909-923
Wagemans J P (1986) Direct theory of perception: an evaluation by representative of indirect theories of perception. Année Psychol 86: 261-273
Walker-Batson D, Barton M M, Wendt J S, Reynolds S (1987) Symbolic and affective non-verbal deficits in left- and right-hemisphere injured adults. Aphasiology 1: 257-262
Warren C, Morton J (1982) The effects of priming on picture recognition. Br J Psychol 73: 117-129
Wechsler A F (1973) The effect of organic brain disease on recall of emotionally charged versus neutral narrative texts. Neurology 23: 130-135
Weintraub S, Mesulam M M, Kramer C (1981) Disturbances in prosody: a right hemisphere contribution to language. Arch Neurol 38: 742-744
Wexler B E (1980) Cerebral laterality and psychiatry: a review of the literature. Am J Psychiatry 137: 279-291
Wolf M E, Goodale M A (1987) Oral asymmetries during verbal and nonverbal movements of the mouth. Neuropsychologia 25: 375-396
Young A W, McWeeny K H, Hay D C, Ellis A W (1986) Matching familiar and unfamiliar faces on identity and expression. Psychol Res 48: 63-68
Zajonc R B (1984) On the primacy of affect. Am Psychol 39: 117-123
Zajonc R B, Markus H (1984) Affect and cognition: the hard interface. In: Izard C E, Kagan J, Zajonc R B (eds) Emotion, cognition, and behavior. Cambridge University Press, Cambridge, pp 73-102
Zenhausern R, Notaro J, Grosso J, Schiano P (1981) The interaction of hemispheric preference, laterality, and sex in the perception of emotional tone and verbal content. Int J Neurosci 13: 121-126
Zillman D (1983) Transfer of excitation in emotional behavior. In: Cacioppo J T, Petty R E (eds) Social psychophysiology: a sourcebook. Guilford, New York, pp 215-240

The Meaning of Emotional Disturbances Resulting from Unilateral Brain Injury

G. Gainotti

Institute of Neurology, 2nd Faculty of Medicine, University of Napoli, Nuovo Policlinico, Via S. Pansini 5, 80131 Napoli, Italy

INTRODUCTION

Although very popular and supported by an extensive body of clinical and experimental research, the hypothesis of a relationship between emotions and hemispheric specialization has not yet found a clear and univocal formulation. In fact, at least four different interpretations have been advanced to account for the diverging emotional changes observed in right and left brain-damaged patients by clinical researchers who first raised the problem of the possible links between emotions and hemispheric division of labor.

The first of these interpretations assumes a general superiority of the right hemisphere for various aspects of emotional behavior (Gainotti 1983, 1988; Ley and Bryden 1981).

The second hypothesis proposes a different specialization of the right and left hemisphere for opposite aspects of the mood, the left underpinning "positive" emotions and the right "negative" affects (Rossi and Rosadini 1967; Sackeim and Gur 1978; Sackeim et al. 1982; Silberman and Weingartner 1986).

The third interpretation posits that the right hemisphere might have a superiority in the field of nonverbal (emotional) communication similar to that shown by the left hemisphere in the realm of propositional language (Ross 1984; Ruckdeschel-Hibbard et al. 1984).

Finally, the last interpretation casts important doubts on the specificity of the laterality effects for emotions observed both in normal subjects and in brain-damaged patients, assuming that these effects might be the byproduct of more basic hemispheric asymmetries concerning language, the processing of complex perceptual material, or the control of subtle facial movements (Gentili 1965; Ekman et al. 1981).

In my opinion, two main reasons can be advanced to explain the difficulties met by researchers who have tried to clarify the links between emotions and hemispheric specialization. The first reason is related to the lack of integration existing in this area between empirical data and theoretical models. Thus, not only was it impossible to insert into a precise theoretical framework the patterns of emotional behavior observed in the clinical studies (Gainotti 1972) which first suggested the hypothesis of a right hemisphere dominance for emotions, but a lack of reference to an explicit theoretical

Experimental Brain Research Series 18

model of emotions can also be observed in the most recent research conducted in this area. A greater integration between theoretical models and empirical data is, therefore, one of the main goals for the future in this field and the present book constitutes one of the first attempts in this direction.

The second reason resides in the very heterogeneous set of situations in which data relevant to the hypothesis of a link between emotions and hemispheric specialization have been gathered. Data relevant to this hypothesis have been obtained, among others from:

1. Clinical and experimental investigations conducted in patients with unilateral brain injury
2. Studies of ictal and interictal emotional behavior of patients with focal epilepsy (usually localized to the temporal lobes)
3. Research conducted in normal subjects or in split-brain patients to investigate the capacity of the intact right and left hemisphere to process emotional information and to produce an appropriate emotional expression
4. Neuropsychological, neuroradiological, and neurophysiological inquiries, conducted in various kinds of psychiatric patients to assess the relationships between hemispheric asymmetries of function and different forms of psychopathology.

Now, data obtained in these heterogeneous clinical situations and with different experimental paradigms have been interpreted according to the theoretical and methodological problems typical of each research field, provoking, in addition to the above-mentioned "general" interpretations, a lot of "local" interpretations, capable only of explaining results obtained in a restricted field of research.

Furthermore, even if we limit ourselves to considering results obtained in a much more homogeneous research field (such as studies dealing with the emotional behavior of patients with unilateral brain injury), we see that different lines of research have provided results consistent with different sorts of interpretations. Table 1 gives a summary, listing the main lines of research followed in the study of the emotional behavior of patients with unilateral brain damage and the consistency of these research with the general interpretations mentioned above.
The data sketched in Table 1 will be analysed in more detail in the next section, where each interpretation will be discussed with reference to results obtained in studies conducted on unilateral brain-damaged patients (since these studies constitute the specific topic of this chapter).

STUDIES OF EMOTIONAL BEHAVIOR IN UNILATERAL BRAIN-DAMAGED PATIENTS: RESULTS OBTAINED FOLLOWING VARIOUS LINES OF RESEARCH AND THEORETICAL INTERPRETATIONS

Hypothesis 1

The hypothesis of a specialization of the right and left hemispheres for opposite aspects of the mood was first advanced by authors who observed a

Table 1. Relationship between results obtained in unilateral brain-damaged patients following different lines of research and interpretation of the links between emotions and hemispheric asymmetries of function

	Main interpretations			
Main lines of research	Specialization of the left hemisphere for positive emotions and of the right hemisphere for negative emotions	Dominance of the right hemisphere for various aspects of emotional behavior	Right hemisphere dominance for nonverbal emotional communication	Emotions are not lateralized; influence of associated variables
Amytal studies	Very consistent	Less consistent	Inconsistent	Inconsistent
Clinical investigations of the emotional changes observed after UBL	Less consistent	Very consistent	Less consistent	Less consistent
Assessment of mood changes following UBL	Consistent	Consistent	Less consistent	Less consistent
Emotional outbursts in patients with UBL	Very consistent	Less consistent	Inconsistent	Inconsistent
Psychophysiological Investigations	Inconsistent	Very consistent	Inconsistent	Inconsistent
Comprehension of emotions and emotional expression	Inconsistent	Very consistent	Very consistent	Partly consistent

UBL, Unilateral brain lesions

different emotional behavior in patients submitted to pharmacological inactivation of the right or left half of the brain (Terzian and Cecotto 1959; Alema' and Donini 1960; Perria et al. 1961; Terzian 1964; Rossi and Rosadini 1967).

These authors reported that injection of sodium amytal into the left carotid artery produces a "depressive-catastrophic" reaction, characterized by a sad and pessimistic attitude, with complaints concerning every aspect of life and with a tendency to weep, whereas pharmacological inactivation of the right hemisphere is followed by an "euphoric-maniacal" reaction, characterized by a relaxed attitude, with an extremely optimistic view of everything and with a tendency to smile, to joke and to break into actual laughter (Rossi and Rosadini 1967).
Since these emotional manifestations were not related to the patients' personality nor to the conditions of examination, they were considered as resulting from disruption of neural mechanisms specifically underpinning opposite aspects of the mood (with a major involvement of the left hemisphere in positive emotions and of the right hemisphere in negative affects).
This hypothesis was later supported and reviewed by several other authors (e.g., Dimond and Farrington 1977; Sackeim and Gur 1978; Ahern and Schwartz 1979; Reuter-Lorenz and Davidson 1981; Davidson 1985; Kinsbourne and Bemporad 1984; Silberman and Weingartner 1986) and has enjoyed great popularity; however, it is not equally supported by all the research lines reported in Table 1. Only one of these lines of research is strongly consistent with the hypothesis, whereas other lines are only in partial agreement with it and still others are clearly inconsistent with the specialization of the left hemisphere for positive emotions and of the right hemisphere for negative affects.

The strongest support to this theory has been provided by a retrospective study of Sackeim et al. (1982) who reviewed the literature on uncontrollable outbursts of laughing and crying and found that in cases of destructive brain pathology outbursts of laughing were associated with predominantly right-sided lesions and outbursts of crying with left-sided injury, whereas in patients with irritative brain lesions the opposite trend was observed. Ictal outbursts of laughing (gelastic epilepsy) were predominantly found in patients with left-sided foci, whereas pathological outbursts of crying (dacristic epilepsy) tended to prevail in patients with right-sided foci.
According to Sackeim et al. (1982) these data demonstrate: (a) that the left hemisphere underpins positive emotions, since excitation of this hemisphere provokes outbursts of laughing with an experience of positive mood, whereas the reverse holds for the right hemisphere; (b) that in patients with unilateral brain damage the overt emotional behavior reflects the prevalent activation of the polar aspect of the mood represented in the intact hemisphere.

Among the lines of research that I consider as only partially consistent with this theory are: (a) clinical investigations which have evaluated the affective and emotional changes resulting from unilateral brain injury by observing the patterns of emotional behavior displayed during a neuropsychological assessment, (b) neuropsychological investigations which have assessed mood changes by means of depression rating scales or of biological markers of depression, such as the dexamethasone suppression test.

Clinical investigations which have analyzed the patterns of emotional behavior shown by right and left brain-damaged patients during neuropsychological examinations (Gainotti 1969, 1972) have, indeed, confirmed that a "catastrophic reaction" follows left-hemisphere injury and that an "indifference reaction" is typical of right-hemisphere damage, but have also considered as misleading the equivalence between "catastrophic reaction" and biological depression and between "indifference reaction" and euphoric-maniacal state. Catastrophic reactions of left brain-damaged patients usually consisted of increasing anxiety reactions or of sudden bursts of tears, triggered by repeated frustrating attempts of verbal communication and observed in a context of severe Broca's aphasia and of motor disability in the right hand. These emotional manifestations were, therefore, considered as a dramatic but psychologically appropriate form of reaction to the catastrophic effects of the disease, rather than as form of biological depression.

On the other hand, right brain-damaged patients with indifference reactions looked in general apathetic and unduly jocular, but also denied their disabilities or expressed exaggerated hatred toward the paralyzed limbs. It seemed very difficult to draw a parallel between these emotional disturbances and the agitated, excited patterns of behavior usually displayed by manic patients.

Also partially consistent with the view of the emotional behavior of left brain-damaged patients as a form of biological depression (and of the emotional reaction of right-sided patients as a sort of euphoric-maniacal state) are investigations which have assessed mood changes resulting from unilateral brain injury by means of depression rating scales or of biological markers of depression.
Some authors (e.g., Black 1975; Gasparrini et al. 1978; Finklestein et al. 1982; Pasoli et al. 1982; Robinson and Price 1982; Robinson et al. 1984) have documented a significant prevalence of depression in left brain-damaged patients, but most (e.g., Dickmen and Reitan 1974, 1977; Black and Black 1982; De Agostini et al. 1983; Bauer et al. 1983; De Bonis et al. 1985; Lipsey et al. 1985; Reding et al. 1985) have failed to find a significant relationship between incidence of depression and laterality of brain injury. Furthermore, some authors (e.g., Folstein et al. 1977; Kulesha et al. 1981; Lim and Ebrahim 1983; Sinyor et al. 1986) have even reported a trend in the opposite direction (see Gainotti 1988 for a more exhaustive discussion of this issue).

Finally, the lines of research that seem more clearly inconsistent with the hypothesis of a different specialization of the right and left hemisphere for opposite aspects of mood are:

1. Psychophysiological investigations which have studied electrodermal activity and pain tolerance in patients with unilateral brain injury;
2. Research which has studied recognition and expression of various types of (positive and negative) emotions in unilateral brain-damaged patients.

Results obtained by psychophysiological studies are in clear contrast to those that are predicted from viewing the emotional change of left brain-damaged patients as a form of biological depression. A number of psychophysiological investigations conducted in patients with endogenous depression have, indeed, shown that the galvanic skin response is significantly reduced in these patients (e.g., Greenfield et al. 1963; Dawson et al. 1977; Storrie et al. 1981; Ward and Doerr 1986), whereas pain tolerance is increased (e.g., Hemphill et al. 1952; Hall and Stride 1954; Merskey 1965; Davis et al. 1979). The same pattern of response should, therefore, be observed in left brain-damaged patients if their catastrophic reaction is to be considered as a form of biological depression. Contrary to this prediction, Heilman et al. (1978); Morrow et al. (1981), and Zoccolotti et al. (1982) have shown that the galvanic skin response is normal in left (and significantly reduced in right) brain-damaged patients, whereas Cubelli et al. (1984) and Neri et al. (1985) have shown that pain tolerance is normal in left (and increased in right) brain-damaged patients.

Even more directly at variance with the hypothesis of specialization of the left and right hemisphere for positive and negative emotions are results of investigations which have studied recognition and expression of various types of emotions in patients with unilateral brain injury.

The hypothesis of hemispheric specialization gives rise to the prediction that left brain-damaged patients, relying mainly on the center for negative emotions in the intact right hemisphere, should be selectively impaired in recognition and expression of positive emotions, whereas the opposite pattern of impairment should be demonstrated by right brain-damaged patients. Results obtained by Cicone et al. (1980), Prigatano and Pribram (1982), Etcoff (1986), Bowers et al. (1985), and Borod et al. (1986) have not been consistent with this hypothesis, since they have generally shown an overall greater impairment of right brain-damaged patients, but no interaction between side of damage and type of emotion to be expressed or recognized.

Hypothesis 2

The hypothesis of a general dominance of the right hemisphere for various aspects of emotional behavior was put forward by Gainotti (1972), who noticed that the theory considering the catastrophic reaction of left brain-damaged patients as a sort of biological depression and the indifference reaction of right-sided patients as a form of manic state was not consistent with the qualitative aspects of the emotional behavior shown by these patients.

Gainotti (1972) proposed, instead, that the catastrophic reaction be considered as a dramatic but psychologically appropriate form of emotional reaction and the indifference reaction of right brain-damaged patients as an abnormal and emotionally inappropriate form of reaction to the disability. According to his viewpoint, left brain-damaged patients have an appropriate (and even a dramatic) emotional reaction because the hemisphere dominant for emotions (namely the right one) is intact. On the contrary, right brain-damaged patients show an abnormal and clearly inappropriate emotional reaction to the disability because in them brain injury involves the right emotionally dominant hemisphere.

In the following years a very large number of clinical and experimental investigations supported the hypothesis of a major involvement of the right hemisphere in various aspects of emotional behavior, so that we can say that of the research lines reported in Table 1 only the observations made during injection of sodium amytal into the right and left carotid artery and the study of pathological outbursts of laughing and crying reported by Sackeim et al. (1982) can be considered as giving results at variance with this hypothesis, whereas all the other lines are more or less consistent.
In particular, highly supportive of this hypothesis are: (a) investigations which have studied comprehension and expression of emotions, and (b) psychophysiological investigations which have studied the electrodermal response and other indices of autonomic activity in patients with unilateral brain injury.

Comprehension and expression of emotions have been studied in brain-damaged patients through the two main channels which are used to communicate emotions, namely facial expression of emotions and affective tone of voice. Most studies conducted along this direction of research have shown a greater impairment of right brain-damaged patients, both at the expressive and at the receptive level, independently of the channel of communication used.

Thus, Cicone et al. (1980), De Kosky et al. (1980), Goldblum (1980), Benowitz et al. (1983), Etcoff (1984), Bowers et al. (1985), and Borod et al. (1986) have shown that patients with right hemisphere injury are seriously impaired in recognizing facial expression of emotions, whereas Heilman et al. (1975). Tucker et al. (1977) and Heilman et al. (1984) have demonstrated that right brain-damaged patients are unable to recognize the emotion expressed by the voice of the speaker, although able to understand the meaning of what he is saying. Analogously, Buck and Duffy (1980) and Borod et al. (1985) have shown that right brain-damaged patients are unable to communicate their emotions through gestures and facial expression of emotions, whereas Tucker et al. (1977), Ross and Mesulam (1979), Hughes et al. (1982), and Borod et al. (1985) have shown that these patients are unable to express emotions through tone of voice.

Psychophysiological investigations which have studied the autonomic response to emotional stimuli of right and left brain-damaged patients have demonstrated that not only the communicative aspects of emotions, but also the more basic functions of emotional arousal are disrupted in patients with right hemisphere injury. These investigations have, indeed, consistently shown that right brain-damaged patients show a defective autonomic response both to painful stimuli (Heilman et al. 1978) and to emotionally laden visual material (Morrow et al. 1981; Zoccolotti et al. 1982).

The hypothesis of a general dominance of the right hemisphere for various aspects of emotional behavior is also strengthened by the criticism that can be addressed to the lines of research which have given results inconsistent with it. Thus, observations made by the first researcher who injected sodium amytal into the right and left carotid arteries have not been confirmed by other who have afterwards used the same procedure (see, for example, Milner 1967; Tsunoda and Oka 1976; Kolb and Milner 1981), whereas the study of uncontrollable outbursts of laughing and crying conducted by Sackeim et al. (1982) suffers the methodological weakness common to all retrospective studies and has not been confirmed by a direct observation of spontaneous facial expressions, occurring at the onset of lateralized focal seizures (Strauss et al. 1983).

Hypothesis 3

The hypothesis assuming that the right hemisphere might have a superiority in the realm of nonverbal communication similar to that shown by the left hemisphere in the control of propositional language stems from both theoretical and empirical reasons.
From the theoretical point of view, this hypothesis draws on Darwin's (1872) claim that communication about emotions, owing to its great biological significance, should be based both on specialized peripheral organs (such as the human facial musculature) and on central nervous system structures specifically subserving comprehension and expression of emotions.

In more recent years the importance of functions of nonverbal communication has been stressed in theories such as the facial feedback hypothesis (Tomkins 1962; see Lanzetta and McHugo, this volume) which assume that affects result from the perception of facial muscular activity, and by the parallel development of sophisticated techniques of analysis of nonverbal communication, such as the FACS procedure developed by Ekman and Friesen (1978) or the PONS test, developed by Rosenthal et al. (1979).

From the empirical point of view, this theory has been prompted by data, reported in the previous section, consistently showing major involvement of the right hemisphere in functions of nonverbal communication.

Drawing on these data, Ross (1981, 1984) and Ruckdeschel-Hibbard et al. (1984) have proposed the existence of a parallelism between the anatomofunctional organization of language at the level of the left hemisphere and the corresponding organization of emotional communication at the level of the right hemisphere. The same authors have also assumed that disruption of emotional communication should be considered as the primary disorder resulting from right hemisphere injury, whereas the other emotional disturbances observed in these patients should be viewed as a consequence of their basic inability to comprehend and express emotions. In particular, the emotional indifference typically observed in these patients, should not be considered as an abnormal form of reaction to the disability, but as the consequence of a primary inability to express an otherwise intact experience of emotions.

If results obtained along the lines of research reported in Table 1 are considered from this point of view, they are not necessarily inconsistent with the above-mentioned hypothesis. Thus, Ross and Rush (1981) have explicitly claimed that right brain-damaged patients are as depressed as left-hemisphere patients but, being unable to communicate their affective state, may give to the examiner or to family members a superficial impression of inappropriate indifference.

It must be acknowledged, however, that, apart from the amytal studies and from the survey of uncontrollable emotional outbursts (which are inconsistent with the hypothesis, but are open to criticism, as I mentioned the end of the last section) this interpretation puts perhaps an excessive stress on disorders of emotional communication of right brain-damaged patients and is, in any case, inconsistent with results of psychophysiological investigations which have studied emotional arousal in patients with unilateral brain injury.

The hypothesis viewing the indifference reaction of right brain-damaged patients as only apparent and resulting from a discrepancy between intact emotional experience and defective emotional expression should, indeed, predict no defect of emotional arousal in these patients, since autonomic arousal faithfully reflects the intensity of an emotional experience (Duffy 1962; Hohmann 1966; Schachter 1975). On the contrary, Heilman et al. (1978), Morrow et al. (1981), and Zoccolotti et al. (1982) have shown that right brain-damaged patients show a defective autonomic response both to painful stimuli and to emotionally loaded visual stimuli, as mentioned when discussing the hypothesis of a general dominance of the right hemisphere for various aspects of emotional behavior.

The caution that we suggest before overstressing defects of emotional communication of right brain-damaged patients proceeds from two reasons. The first (which will be more extensively discussed in the next section) concerns the specificity of these defects, since, according to some authors, the right hemisphere advantage in discriminating emotional stimuli could derive from the sensory characteristics of the stimuli and not from their emotional nature. The second reason consists in the fact that some authors have failed to confirm the greater impairment of right brain-damaged patients on tasks of emotional communication. Negative results have been reported, for example, by Schlanger et al. (1976) and by Prigatano and Pribram (1982) studying the recognition of emotionally laden stimuli and by Kolb and Milner (1981) studying the facial emotional expressiveness of right and left brain-damaged patients.

Hypothesis 4

The last hypothesis maintains that emotions are not lateralized and that laterality effects usually attributed to emotions are actually due to the influence of other lateralized functions (such as language or visual-spatial analysis) which can have a nonspecific influence on results obtained on "emotional" tasks.

The first proponent of this theory was, perhaps, Gentili (1965) who maintained that right brain-damaged patients are not really indifferent, but simply unaware of their disability, since they pay no attention to the left (paralyzed) side of the body. According to this authors, the examiner who observes from the outside the amazing lack of concern for the disability shown by these patients, attributes to emotional indifference a pattern of behavior which is basically due to a severe neglect for the left half of the body and of extrapersonal space.

In the following years, similar interpretations have been advanced to explain laterality effects for emotions observed both in normal subjects and in brain-damaged patients. The general question raised was whether poor results obtained by right brain-damaged patients on tasks of emotional comprehension and expression proceed from a specific impairment in the processing of emotional material or can be simply attributed to a more general impairment in the processing of complex perceptual material.
Some results have suggested that the latter could be the case. Thus, De Kosky et al. (1980), contrasting results obtained by right and left brain-damaged patients on tasks requiring the identification of emotional visual stimuli and of nonemotional neutral faces, showed that right hemisphere patients scored significantly worse than left-sided patients on various emotional tasks, but that the intergroup differences disappeared when a subsequent analysis of variance was carried out using as covariate the scores obtained on the test of neutral faces identification.

Analogously, Weintraub et al. (1981) have suggested that the right brain-damaged patients' impairment in recognizing the affective components of speech might be nonspecific, since the same patients also show a more general deficit in processing the nonaffective aspects of speech prosody.

In my opinion, the influence of nonspecific factors on results obtained on "emotional" tasks must be carefully controlled in any research on this topic, but the importance of the problem should not be overemphasized. The indifference reaction of right brain-damaged patients, for example, cannot be simply considered as a byproduct of inattention for the left side of the body and of extrapersonal space; as early as 1914 Babinski had made a careful distinction between "anosognosia" and "anosodiaphoria", applying the first label to patients who were seemingly unaware of their left-sided hemiplegia and the second to patients who were conscious of their severe disability, but treated it with undue unconcern and cheerful acceptance.

Since most right brain-damaged patients showing a classical indifference reaction are fully aware of their disability, the interpretation of the emotional indifference proposed by Gentili (1965) might, perhaps, account for the reactions of a few of these patients, but not for most of them. Analogously, although nonspecific factors may partly account for defects of emotional communication shown by right brain-damaged patients, some studies (e.g., Hughes et al. 1982; Etcoff 1986; Bowers et al. 1985) have shown that disorders in recognition of nonemotional prosody and of neutral faces are partly independent of defective identification of emotional prosody and of emotional facial expression.

These findings suggest that processing of nonemotional complex perceptual material and processing of emotional information are subserved by independent structures located in different parts of the right hemisphere

A NEW RESEARCH STRATEGY IN PATIENTS WITH UNILATERAL BRAIN INJURY TO CLARIFY THE RELATIONSHIPS BETWEEN EMOTIONS AND HEMISPHERIC SPECIALIZATION

Since results obtained along the lines of research reported in Table 1 had not allowed a clear and unequivocal interpretation of the links existing between emotions and hemispheric specialization, we have recently tried, in cooperation with Pizzamiglio, Caltagirone, Zoccolotti, Mammucari, Ekman, Friesen and Guariglia, to tackle this problem using a new research strategy.

The main characteristic which differentiates the new research strategy from those followed in previous investigations concerns the complexity of the experimental design, inasmuch as previous investigations had generally taken one aspect of the emotional behavior of patients with unilateral brain injury into account at a time, whereas our research program tried simultaneously to evaluate various aspects of the emotional behavior of right and left brain-damaged patients A similar research strategy has been recently adopted by other authors (e.g., Etcoff 1986 and this volume; Borod et al. 1985, 1986) since studying simultaneously in the same patients different aspects of emotional behavior can offor tho following important advantages:

1. The degree of consistency existing between various aspects of emotions usually disrupted by right brain injury can be determined (e.g., emotional indifference, poor comprehension and expression of emotions, and defects of emotional arousal)
2. It can be checked whether all these aspects of emotional behavior are equally sensitive to the presence of a right hemisphere lesion or whether some of them are more frequently and more strongly affected
3. It can be determined whether disorders observed in different emotional domains are specific or can be considered as the byproduct of other more basic disturbances resulting from right hemisphere damage.

Coming back from the general characteristics and advantages of the new research strategy to the specific purposes of our research program, we can say that our investigation aimed to study simultaneously the following components of the emotional behavior of right and left brain-damaged patients: (a) emotional reaction to the disability; (b) comprehension of emotional stimuli; (c) facial expression of emotions; (d) autonomic response to emotional stimuli.

The quality of the emotional reaction to the disability was evaluated clinically, observing the emotional behavior shown by the patient during a standard neuropsychological examination and checking the reliability of this clinical impression against the evaluation of the same phenomenon given by nurses and by family members. This aspect of our research has basically confirmed previous clinical data, but will not be specifically taken into account in the present report.

Comprehension of emotional stimuli was assessed by administering to the patients a set of photographs of facial emotional expressions kindly provided by Ekman and Friesen (1975). This set consists of 36 photographs of six actors expressing each the six basic emotions of happiness, fear, anger, disgust, surprise, and sadness. For each item of the test, patients were requested to select from among four emotional faces the one expressing the same emotion depicted in the target face. In order to determine whether performances obtained on this task were due to the emotional nature of the stimuli or to their physical characteristics, patients also received the face recognition test of Benton and Van Allen (1968), which is very similar to the facial emotion recognition test, but differs in that it requires identification of facial identity and not of facial emotional expression.

Furthermore, in order to minimize the influence of unilateral spatial neglect on scores obtained by right brain-damaged patients both on the face recognition and on the facial emotion recognition test, target face and multiple choice alternatives were in any case arranged in a vertical array in the middle or the sheet.

Facial expression of emotions was studied by showing the patients through a rear projector movies devised to elicit pleasant, neutral, and strongly unpleasant emotional reactions and recording with a videotape their facial expressions. In order to determine whether any possible intergroup differences in facial emotional expression were due to the emotional nature of the task or to other hemispheric asymmetries concerning the control of facial movements, each patient also received a test of imitation of nonemotional facial movements.

More detailed information about the procedures used to elicit emotions and the scoring system used to evaluate emotional expressiveness is reported in the chapter by Pizzamiglio and Mammucari (this volume), which specifically reports results obtained in this part of our research program.

Autonomic responsivity to emotional stimuli was studied by recording the galvanic skin response and heart rate changes during the projection of films used to elicit positive, neutral, and negative emotional reactions. Electrodermal response was taken as an index of sympathetic arousal, whereas heart rate deceleration was retained as an index of parasympathetic

arousal. For more detailed information concerning procedures used to study the autonomic responsivity, the reader is referred to the chapter by Caltagirone et al. (this volume) which is specifically devoted to discussion of the results obtained in this part of our research.
The following general outcomes were predicted from the main theories advanced to explain the links between emotions and hemispheric specialization:

If the left hemisphere subserves positive emotions and the right hemisphere subserves negative emotions, then an interaction between laterality of lesion and type of emotion should be observed in the various sections of our research program. Left brain-damaged patients should be more impaired in recognition, expression, and autonomic response to positive emotions, whereas just the opposite pattern of impairment should be shown by right-hemisphere patients.

If the right hemisphere is more critically involved in all the aspects of emotional behavior, then right brain-damaged patients should demonstrate:
1. A greater incidence of emotional inappropriateness
2. A greater impairment on tests of emotional comprehension and of emotional expression
3. A reduced level of emotional arousal

If disorders of emotional behavior shown by right brain-damaged patients derive from a basic inability to comprehend emotional information and to express an otherwise intact emotional experience, then these patients should be particularly impaired on tasks of emotional comprehension and expression, but demonstrate a normal autonomic reactivity.

If the greater "emotional impairment" of right brain-damaged patients is simply the byproduct of nonspecific factors (such as poor visual-spatial analysis or defective control of subtle facial movements), then right brain-damaged patients should demonstrate a similar impairment in the comprehension of emotional and of nonemotional stimuli and in the production of emotional and nonemotional facial movements. No impairment should, on the contrary, be observed in the autonomic response to emotional stimuli.

GENERAL SURVEY OF THE MAIN RESULTS OBTAINED IN OUR RESEARCH PROGAM

In this report only a general survey of the main results obtained in our research program will be provided, as in other chapters of this volume Pizzamiglio and Mammucari more analytically present data concerning the facial expression of emotions, whereas Caltagirone et al. exhaustively discuss data concerning the autonomic reactivity. A first general claim that can be made is that unexpected results were obtained on the emotional communication tasks, in that no significant difference was found between right and left brain-damaged patients on tests of emotional comprehension and expression, although in both situations control subjects scored significantly better that either group.

Table 2. Mean scores obtained by normal controls and right and left brain-damaged patients on the test of neutral faces identification and of emotional facial expressions recognition

	Normal control	Right brain-damaged patients	Left brain-damaged patients
Neutral faces Identification	42.37 (5.21)	38.50 (6.92)	38.90 (6.35)
Recognition of emotional facial expression	15.21 (2.69)	11.43 (4.34)	11.10 (3.45)

Data obtained by Guariglia (1986) and reported in Table 2 have shown that no significant difference exists between right and left brain-damaged patients on tests of recognition of facial emotional expression (Ekman and Friesen 1975) and of identification of neutral faces (Benton and Van Allen 1968).

The contrast between these findings and most other results reported in the literature cannot be easily explained, but it is possible that part of this discrepancy may be due to the fact that in our study we have carefully ruled out the influence of unilateral spatial neglect, which has not always been controlled in other investigations.

The control of this variable could be important because some authors (e.g., Goldblum 1980; De Kosky et al. 1980) have shown that unilateral spatial neglect and impaired recognition of emotional faces are significantly related in right brain-damaged patients. It is well known, however, that in matching to sample tasks using spatially distributed response alternatives, unilateral neglect may spuriously decrease scores obtained by right brain-damaged patients (Gainotti et al. 1977). Having placed the stimulus face and the response alternatives in a vertical column could, therefore, have contributed to reducing the difference between right and left hemisphere patients.

Results similar to those obtained by Guariglia (1986) at the receptive level have been obtained by Mammucari et al. (1988) at the expressive level and are reported in this volume by Pizzamiglio and Mammucari. These authors observed no significant difference between patients with right-sided and left-sided injury when they analyzed the frequency with which these subjects displayed facial expressions congruent to emotional movies. In this case too, control subjects produced a significantly higher number of congruent facial emotional expressions than either hemispheric group.

In the same experimental situation, however, an interesting difference was observed between right hemisphere patients and the other control groups during the unpleasant emotional film, which consisted of a bloody presentation of a surgical toilette. In this situation, normal controls and left brain-damaged patients showed a tendency to move their eyes away from the screen, whereas right brain-damaged patients did not show a similar pattern of behavior. This lack of avoidance eye movements suggests an emotional indifference of patients with right brain injury in face of stressing situations.

Results obtained studying the autonomic response to emotional stimuli are well in line with this interpretation. Zoccolotti et al. (1986) have shown, indeed, that the galvanic skin response shown by right brain-damaged patients during the emotional movies was significantly different from that shown by normal controls and by left-hemisphere damaged patients. The former presented a very weak electrodermal response, which showed a further progressive decline during the course of the experimental session without being influenced by the emotional or nonemotional nature of the stimuli (as data reported in Fig. 1 clearly show). The latter, on the contrary, presented a much more specific reaction to the emotional stimuli, showing the highest level of response during the projection of the emotional (and above all of the unpleasant) visual scenes.

Similar conclusions were reached when attention was focused on heart rate deceleration, considered as an index of parasympathetic activation. In this case, too, normal controls and left brain-damaged patients showed the greatest heart rate deceleration while viewing the unpleasant emotional film, whereas right hemisphere patients did not significantly change their heart rate as a function of the emotional or nonemotional and of the positive or negative emotional valence of the visual stimuli.
In conclusion, our results do not support the idea of specialization of the left hemisphere for positive affects and of the right hemisphere for positive emotions, inasmuch as no interaction was observed between laterality of lesion and type of emotion in recognition, expression, and autonomic response to emotional stimuli.

Our results are also at variance with the theory viewing disruption of emotional communication as the primary disorder resulting from right hemisphere damage, since no significant difference was found between right-sided and left-sided patients on tasks of emotional comprehension and of facial emotional expression, whereas a reduced level of autonomic emotional arousal was observed in right brain-damaged patients.

On the contrary, our results do support the theory considering the emotional reaction of right brain-damaged patients as abnormal and inappropriately indifferent, since these patients did not move the eyes away from very unpleasant visual scenes (as normal controls and left-hemisphere patients usually did) and did not show in face of emotional stimuli the pattern of autonomic reactivity presented by normal controls and by patients with left-hemisphere injury.

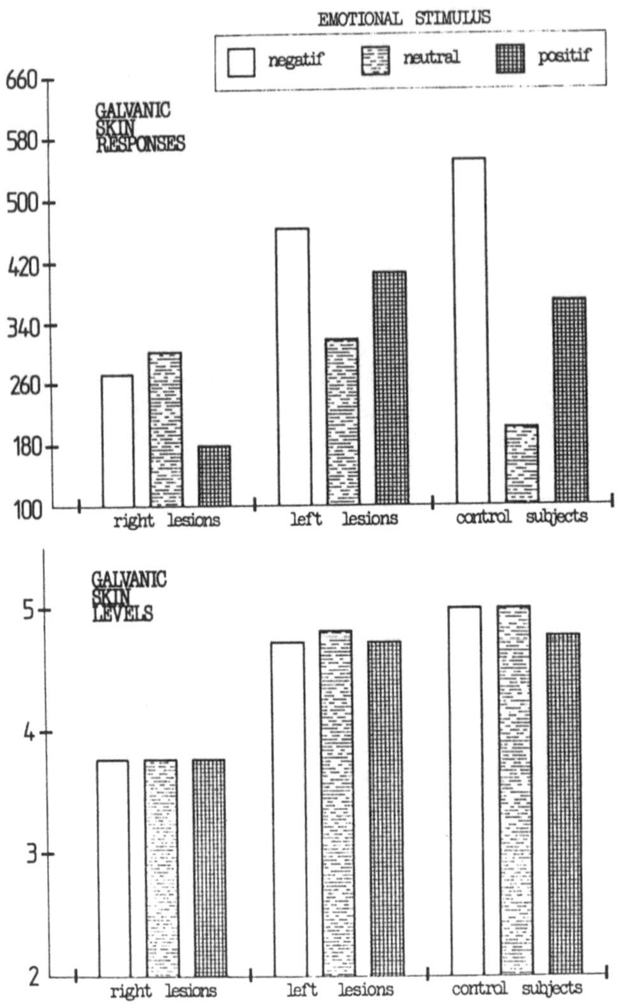

Fig 1. Galvanic skin levels and galvanic skin responses shown by right and left brain-damaged patients and by control subjects during the neutral and the emotional movies

Thus, although our findings do not fully support any of the four theories advanced to explain the relationships between emotions and hemispheric specialization, they do confirm the critical role played by the right hemisphere in emotional behavior. Furthermore, they seem to suggest that the right hemisphere dominance concerns more the inner experience of emotions and the concomitant patterns of autonomic arousal than the comprehension of emotional stimuli and the expressiveness of the emotional response.

REFERENCES

Ahern GL, Schwartz GE (1979) Differential lateralization for positive versus negative emotion. Neuropsychologia 17: 693-698

Alemà G, Donini G (1960) Sulle modificazioni cliniche ed elettroencefalografiche da introduzione intracarotidea di iso-amil-etil-barbiturato di sodio nell'uomo. Boll Soc Ital Biol Sper 36: 900-904

Babinski J (1914) Contribution à l'étude des troubles mentaux dans l'hemiplegie organique cerebrale (Anosognosie). Rev Neurol (Paris) 27: 845-848

Bauer M, Gans J, Harley J (1983) Dexamethasone suppression test and depression in a rehabilitation setting. Arch Phys Med Rehabil 64: 421-422

Benowitz LI, Bear DM, Mesulam MM, Rosenthal R, Zaidel EE, Sperry RW (1983) Nonverbal sensitivity following lateralized cerebral injury. Cortex 19: 5-12

Benton AL, Van Allen MW (1968) Impairment in facial recognition in patients with cerebral disease. Cortex 4: 344-358

Black FW (1975) Unilateral brain lesions and MMPI performance: a preliminary study. Percept Mot Skills 40: 87-93

Black FW, Black IL (1982) Anterior-posterior locus of lesion and personality: support for the caudality hypothesis. J Clin Psychol 38: 3, 468-477

Borod JC, Koff E, Perlman M, Nicholas M (1985) Channels of emotional expression in patients with unilateral brain damage. Arch Neurol 42: 345-348

Borod JC, Koff E, Perlman-Lorch M, Nicholas M (1986) The expression and perception of facial emotion in brain-damaged patients. Neuropsychologia 24: 169-180

Bowers D, Bauer RM, Coslett HB, Heilman KM (1985) Processing of faces by patients with unilateral hemisphere lesions. Dissociation between judgments of facial affect and facial identity. Brain Cogn 4: 258-272

Buck R, Duffy RJ (1980) Nonverbal communication of affect in brain-damaged patients. Cortex 16: 351-362

Cicone M, Wapner W, Gardner H (1980) Sensitivity to emotional expressions and situations in organic patients. Cortex 16: 145-158

Cubelli R, Caselli M, Neri M (1984) Pain endurance in unilateral cerebral lesions. Cortex 20: 369-375

Darwin C (1872) The expression of emotion in man and animals. John Murray, London (Current edition, University of Chicago Press, Chicago 1986)

Davidson RJ (1985) Hemispheric specialization for cognition and affect. In: Gale A, Edwards J (eds) Physiological correlates of human behavior. Academic, London

Davis GC, Buchsbaum MS, Bunney WE Jr (1979) Analgesia to painful stimuli in affective illness. Am J Psychiatry 136: 1148-1151

Dawson ME, Schell AM, Catania JJ (1977) Autonomic correlates of depression and clinical improvement following electroconvulsive shock therapy. Psychophysiology 14: 569-577

De Agostini I, Blanc-Garin J, De Bonis M (1983) Evaluation du comportment émotionnel chez les malades avec lesions cerebrales. Premiers resultats. C R Congr Psychiatrie Neurol Langue Francaise 81: 321-328

De Bonis M, Dellatolas G, Rondot P (1985) Mood disorders in left and right brain-damaged patients: comparison between ratings and self-ratings on the same adjective mood scale. Psychopathology 18: 286-292

De Kosky ST, Heilman KM, Bowers D, Valenstein E (1980) Recognition and discrimination of emotional faces and pictures. Brain Lang 9: 206-214

Dikmen S, Reitan R (1974) MMPI correlates of localized cerebral lesions. Percept Mot Skills 39: 831-840

Dikmen S, Reitan R (1977) MMPI correlates of adaptive ability deficits in patients with brain lesions. J Nerv Ment Dis 165: 247-254

Dimond SJ, Farrington J (1977) Emotional response to films shown to the right or left hemisphere of the brain, measured by heart rate. Acta Psychol (Amst) 4: 255-260

Duffy E (1962) Activation and behavior. Wiley, New York

Ekman P, Friesen WV (1975) Unmasking the face. Prentice-Hall, Englewood Cliffs, NJ

Ekman P, Friesen WV (1978) The facial action coding system. Consulting Psychologists Press, Palo Alto, CA

Ekman P, Hager J, Friesen WV (1981) The symmetry of emotional and deliberate facial action. Psychophysiology 18: 101-106

Etcoff N (1986) The neuropsychology of emotional expression. In: Goldstein G, Tarter RE (eds) Advances in clinical neuropsychology. Plenum Press, New York

Finklestein S, Benowitz LI, Baldessarini RJ, Arana GW, Levine D, Woo E, Bear D, Moya K, Stoll AL (1982) Mood, vegetative disturbance, and dexamethasone suppression test after stroke. Ann Neurol 12: 463-468

Folstein MF, Maiberger R, McHugh PR (1977) Mood disorder as a specific complication of stroke. J Neurol Neurosurg Psychiatry 40: 1018-1020

Gainotti G (1969) Reactions "catastrophiques" et manifestation d'indifference au cours des atteintes cerebrales. Neuropsychologia 7: 195-204

Gainotti G (1972) Emotional behavior and hemispheric side of lesion. Cortex 8: 41-55

Gainotti G (1983) Laterality of affect: the emotional behavior of right and left brain-damaged patients. In: Myslobodsky MS (ed) Hemisyndromes: psychobiology, neurology, psychiatry. Academic, New York

Gainotti G (1988) Disorders of emotions and affects in patients with unilateral brain damage. In: Boller F, Grafman J (eds) Handbook of neuropsychology. Elsevier, Amsterdam

Gainotti G, Caltagirone C, Miceli G (1977) Poor performances of right brain-damaged patients on Raven's Coloured Matrices. Neuropsychologia 15: 675-680

Gasparrini WG, Satz P, Heilman KM, Goolidge FL (1978) Hemispheric asymmetries of affective processing as determined by the Minnesota Multiphasic Personality Inventory. J Neurol Neurosurg Psychiatry 41: 470-473

Gentili C (1965) Corpo come relazione. Sistema Nerv 17: 174-187

Goldblum MD (1980) La reconnaissance des expressions faciales émotionnelles et conventionnelles au course de lésions corticales. Rev Neurol (Paris) 136: 711-719

Greenfield N, Katz D, Alexander A, Roessler R (1963) The relationship between physiological and psychological responsivity: depression and galvanic skin response. J Nerv Ment Dis 136: 535-539
Guariglia C (1986) Riconoscimento delle espressioni facciali emozionali in soggetti cerebrolesi. Thesis, University of Rome
Hall KRL Stride (1954) The varying response to pain in psychiatric disorders. A study in abnormal psychology. Br J Med Psychol 27: 48-60
Heilman KM, Scholes R, Watson RT (1975) Auditory affective agnosia: disturbed comprehension of affective speech. J Neurol Neurosurg Psychiatry 38: 69-72
Heilman KM, Schwartz HD, Watson RT (1978) Hypoarousal in patients with neglect and emotional indifference. Neurology 28: 229-232
Heilman KM, Bowers D, Speedie L, Coslett HB (1984) Comprehension of affective and nonaffective prosody. Neurology 34: 917-921
Hemphill RE, Hall KRL, Crookes GG (1952) A preliminary report on fatigue and tolerance in depressive and psychoneurotic patients. J Ment Sci 98: 433-440
Hohmann G (1966) Some effects of spinal cord lesions on experimental emotional feelings. Psychophysiology 3: 143-156
Hughes CP, Chan JL, Su MS (1982) Aprosodia in Chinese patients with right cerebral hemisphere lesions. Arch Neurol 40: 732-736
Kinsbourne M, Bemporad B (1984) Lateralization of emotion: a model and the evidence. In: Fox NA, Davidson RJ (eds) The psychobiology of affective development. Erlbaum, Hillsdale NJ
Kolb B, Milner B (1981) Observations on spontaneous facial expression after focal cerebral excisions and after intracarotid injection of sodium amytal. Neuropsychologia 19: 505-514
Kulesha D, Moldosky H, Urowitz M, Zeman R (1981) Brain scan lateralization and psychiatric symptoms in systemic lupus erythematosus. Biol Psychol 16: 407-411
Ley RG, Bryden HP (1981) Consciousness, emotion and the right hemisphere. In: Underwood G, Stevens R (eds) Aspects of consciousness, vol 2, Structural issues. Academic, London
Lim ML, Ebrahim SBJ (1983) Depression after stroke: a hospital treatment survey. Postgrad Med 59: 489-491
Lipsey JR, Robinson RG, Pearlson GD, Rao K, Price TR (1985) The dexamethasone suppression test and mood following stroke. Am J Psychol 42: 317-323
Mammucari A, Caltagirone C, Ekman P, Friesen W, Gainotti G, Pizzamiglio L, Zoccolotti P (1988) Spontaneous facial expression of emotions in brain-damaged patients. Cortex (in press)
Merskey H (1965) The effect of chronic pain upon the response to noxious stimuli by psychiatric patients. J Psychiatr Res 8: 405-419
Milner B (1967) Brain mechanisms suggested by studies of the temporal lobes. In: Milikan CH, Darley FL (eds) Brain mechanisms underlying speech and language. Grune and Stratton, New York
Morrow L, Vrtunski B, Kim Y, Boller F (1981) Arousal responses to emotional stimuli and laterality of lesion. Neuropsychologia 19: 65-71
Neri M, Vecchi GP, Caselli M (1985) Pain measurements in right-left cerebral lesions. Neuropsychologia 23: 123-126

Pasoli EA, Trabucco G, Pasoli CA (1982) Asimmetrie emotive e lesioni cerebrali lateralizzate. G It Psicol 9: 2, 309-322
Perria L, Rosadini G, Rossi GF (1961) Determination of side of cerebral dominance with Amobarbitol. Arch Neurol 4: 173-181
Prigatano GP, Pribram KH (1982) Perception and memory of facial affect following brain injury. Percept Mot Sklills 54: 859-869
Reding M, Orto L, Willensky P, Fortuna I (1985) The dexamethasone suppression test. An indicator of depression in stroke but not a predictor of rehabilitation outcome. Arch Neurol 42: 209-212
Reuter-Lorenz P, Davidson RJ (1981) Differential contribution of the two cerebral hemispheres to the perception of happy and sad faces. Neuropsychologia 19:: 609-613
Robinson RG, Kubos KL, Starr LB, Rao K, Price TR (1984) Mood disorders in stroke patients: importance of lesion location. Brain 107: 81-93
Robinson RG, Price TR (1982) Post-stroke depressive disorders: a follow-up study of 103 outpatients. Stroke 13: 635-641
Rosenthal R, Hall JA, Archer D, Di Matteo M, Rogers PL (1979) The PONS test: measuring sensitivity to nonverbal cues. In: Weitz S (ed) Nonverbal communication. Oxford University Press, New York, pp 357-370
Ross ED (1981) The aprosodias: functional-anatomic organization of the affective components of language in the right hemisphere. Arch Neurol 38: 561-569
Ross ED (1984) Right hemisphere's role in language, affective behavior and emotion. Trends Neurosci 7: 342-346
Ross ED, Mesulam M (1979) Dominant language functions in the right hemisphere? Arch Neurol 36: 144-148
Ross ED, Rush AJ (1981) Diagnosis and neuroanatomical correlates of depression in brain-damaged patients. Arch Gen Psychiatry 38: 1344-1354
Rossi GF, Rosadini G (1967) Experimental analysis of cerebral dominance in man. In: Millikan CJ, Darley FL (eds) Brain mechanisms underlying speech and language. Grune and Stratton, New York, pp 167-174
Ruckdeschel-Hibbard M, Gordon WA, Diller L (1984) Affective disturbances associated with brain damage. In: Filskov S, Boll T (eds) Handbook of neuropsychology, vol 2. Wiley, New York
Sackeim A, Greenberg MS, Weiman L, Gur RC, Hungerbuhler JP, Geschwind N (1982) Hemispheric asymmetry in the expression of positive and negative emotions. Arch Neurol 39: 210-218
Sackeim HA, Gur RC (1978) Lateral asymmetry in intensity of emotional expression. Neuropsychologia 16: 473-481
Schachter S (1975) Cognition and peripheralist-centralist controversies in motivation and emotion. In: Gazzaniga MS, Blakemore C (eds) Handbook of psychobiology. Academic, New York, pp 529-564
Schlanger BB, Schlanger P, Gerstman LJ (1976) The perception of emotionally toned sentences by right hemisphere-damaged and aphasic subjects. Brain Lang 3: 396-403
Silberman EK, Weingartner H (1986) Hemispheric lateralization of functions related to emotion. Brain Cogn 5: 322-353
Sinyor D, Jacques P, Kaloupek DG, Becker R, Goldenberg M, Coopersmith H (1986) Poststroke depression and lesion location. Brain 109: 537-546

Storrie MC, Doerr HO, Johnson MH (19891) Skin conductance characteristics of depressed subjects before and after therapeutic intervention. J Nerv Ment Dis 69: 3, 176-179

Strauss E, Wada J, Kosaka B (1983) Spontaneous facial expressions occurring at onset of focal seizure activity. Arch Neurol 40: 545-547

Terzian H (1964) Behavioural and EEG effects of intracarotid sodium amytal injection. Acta Neurochir (Wien) 12: 230-239

Terzian H, Cecotto C (1959) Determinazione e studio della dominanza emisferica mediante iniezione intracarotide di Amytal sodico nell'uomo. I. Modificazioni Cliniche. Boll Soc Ital Biol Sper 35: 1623-1626

Tomkins SS (1962) Affect, imagery and consciousness. Springer Publ, New York

Tsunoda T, Oka M (1976) Lateralization for emotion in the human brain and auditory cerebral dominance. Proc Jap Acad 52: 528-531

Tucker DM, Watson RT, Heilman KM (1977) Discrimination and evocation of facial recognition in patients with unilateral cerebral lesions. Cortex 3: 317-326

Ward NG, Doerr HO (1986) Skin conductance. A potentially sensitive and specific marker for depression. J Nerv Ment Dis 174: 553-559

Weintraub S, Mesulam MM, Kramer L (1981) Disturbances in prosody: a right hemisphere contribution to language. Arch Neurol 38: 742-744

Zoccolotti P, Scabini D, Violani V (1982) Electrodermal responses in patients with unilateral brain damage. J Clin Neuropsychol 4: 143-150

Zoccolotti P, Caltagirone C, Benedetti N, Gainotti G (1986) Perturbation des reponses vegétatives aux stimuli émotionnels au cours des lesions hemispheriques unilaterales. Encéphale (Paris) 12: 263-268

Recognition of Emotions in Patients with Unilateral Brain Damage

N. L. Etcoff

Massachusetts Institute of Technology, Dept. of Brain and Cognitive Sciences, E10-014, Cambridge, Massachusetts 02139, USA

INTRODUCTION

In this paper, I will present research that I have been conducting on the recognition of facial expressions of emotion in both normal and brain-damaged adults. This research has used facial expression as a vehicle by which to address several important questions about the neuropsychology of emotion.

The first question has to do with the autonomy or modularity of emotion. Simply put, can we distinguish "emotional" functions from general cognition? The particular question addressed here is whether facial expressions of emotion can be recognized separably from the identity of the person displaying the emotion. Answers to questions such as this are of interest both to the emotion theorist who wishes to understand the role of more general cognitive processes in emotional functioning, and to the neuropsychologist who wishes to argue for a biologically autonomous or isolable system.
If the systems governing emotion are not at least partially specialized there can be no biology or psychology of emotion separate from, for example, the biology or psychology of perception in general or motor control in general.

The second question is this: are at least some emotional functions cortically represented and lateralized? More specifically, are emotional functions represented in the right cerebral hemisphere? When one thinks of the biological mediation of emotion one tends to think of older, subcortical brain structures. However, the increasing refinement and elaboration of emotional expression and behavior may have necessitated cortical mediation of many functions. The particular question addressed here is whether there is cortical mediation of the ability to recognize facial expressions of emotion. Embedded within this, is the question of whether or not the division of facial expressions into those that convey negative versus positive emotions is a crucial or even useful one for the study of emotion recognition.

Finally, there is the question of the relations among the various emotional functions such as perceiving emotion, displaying emotion, or showing appropriate emotional response. One may ask whether these emotional functions are best thought of as highly parasitic on one another and virtually

Experimental Brain Research Series 18

indivisible or as more or less autonomous. The particular question addressed in the research to be discussed here is whether the ability to accurately perceive emotional expression in others correlates with the ability to show appropriate emotional concern about one's own well being.

Modularity of Emotion

We know that human beings are capable simply by contracting particular facial muscles of effectively conveying happiness, surprise, fear, anger, disgust, and sadness. There is also evidence for the universality and the biological evolution of these facial expressions: the particular facial muscles recruited for the signaling of each of these emotional states is the same across widely divergent cultures (see e.g., Darwin 1872; Ekman and Friesen 1971). If particular muscles are systematically recruited for the expression of particular emotional states, and if the contractions of these muscles leads to universally recognized facial configurations, a specialized system may have evolved which was capable of discriminating these expressions and making inferences about emotional states from them.

There is considerable evidence that facial expressions are, on the whole, better recognized by the right cerebral hemisphere. This is a reasonably robust finding which emerges both when normal adults are tested through the divided visual field method (e.g., Landis et al. 1979; Ley and Bryden 1979; Strauss and Moscovitch 1981) or when adults with damage to the right versus left hemisphere are compared (e.g., Cicone et al. 1980; Dekosky et al. 1980; Benowitz et al. 1983). There is also evidence that faces and complex visuospatial patterns in general are better recognized by the right hemisphere (e.g., De Renzi and Spinnler 1966; Warrington and James 1967; Rizzolatti et al. 1971; Gazzaniga and Smylie 1983). The question facing us is whether the ability to recognize facial expression can be dissociated from the other visuospatial abilities subserved by the right hemisphere.

Previous research has provided a tentative yes in response to this question. There is evidence from the literature on prosopagnosics, people who have difficulty recognizing familiar faces
Prosopagnosics know that what they are seeing is a face but cannot tell you that the face belongs to a person they should know well, such as their spouses or even themselves. Nonetheless, there are reports that they can recognize and attach meaning to facial expressions of emotion (Shuttleworth et al. 1982; Bruyer et al. 1983). Another source of evidence comes from studies of electrical stimulation mapping of the brain (Fried et al. 1982). When discrete cortical sites are stimulated in patients undergoing right-sided intracranial operations for the treatment of epilepsy, the recognition of facial expressions is disrupted by stimulation of the middle temporal gyrus while the ability to match and remember faces is disrupted by stimulation of other regions (in the superior temporal gyrus, inferior frontal gyrus, and parieto-occipital junction). Interestingly, although there is some variability among subjects in the exact localization of these functions, none showed an overlap

between the areas mediating recognition of the identity of a face and recognition of its emotional state.

Although this evidence is intriguing, its interpretation is not as clear-cut as one would like. For example, it is not known how much the brain organization of an individual suffering from severe, intractable epilepsy reflects that of a normal brain, especially since the focus of the epilepsy is in the hemisphere under scrutiny. Prosopagnosics are rare, and reports of their deficits outside of face recognition are often not presented in great detail.

More systematic studies of larger groups of normal subjects or of patients with unilateral brain damage have generally supported the idea of independence of perceiving identity and emotion (e.g., Cicone et al. 1980; Strauss and Moscovitch 1981; Pizzamiglio et al. 1983). However, at least one study has provided evidence that the recognition of faces and facial expressions may not be dissociable (Hansch and Pirozzolo 1980) while others have suggested the possibility of interdependence, with either the ability to recognize faces dependent on the ability to recognize facial expressions (Ley and Bryden 1979) or the ability to recognize facial expressions dependent on the ability to recognize faces (Dekosky et al. 1980).

One source of confusion in this literature is that arguments for or against independence have often involved comparisons of performance on separate tests of face recognition and facial expression recognition. The problem with such an argument is that if the range or the discriminability of the sets used in testing these abilities is not approximately equal, differences in performance could simply reflect differences in the difficulty of the particular tests, not differences in people's intrinsic ability to perform the two types of discriminations. The core problem is that we do not have a metric that could be applied both to a set of individual faces and a set of individual expressions which could ensure that the exemplars of each type chosen in an experiment are equally representative of the intrinsic difficulty of that type of discrimination in general.

One way around this problem is to provide subjects with a task in which differences in patterns of performance can be demonstrated in the same subjects when faced with tasks involving the discrimination of faces and of facial expressions, rather than simply differences in absolute performance. I devised such a task (Etcoff 1983, 1984a) by adapting techniques used by Wendell Garner to study the ability to attend selectively to a single physical dimension when stimuli vary along two or more dimensions (Garner 1974, 1976).

In this technique, to test whether people can attend selectively to a dimension such as the color of a figure (in this example, let us say red vs blue) while ignoring its shape (circle vs square), subjects sort a deck of cards into two piles. They are presented these cards under three conditions: a "correlated" condition in which the two dimensions covary (for example, the deck consists of only blue squares and red circles); a "constant" condition in which one dimension remains at a constant value while the other varies (for example, the deck consists only of red circles and squares; or only of blue and red circles); and an "orthogonal" condition in which both attributes vary randomly (the deck includes red circles, blue circles, red squares, and blue squares). In each case subjects must sort the stimuli into two piles on the basis of one of the dimensions. If subjects do so equally quickly and accurately in all three conditions, one concludes that the two dimensions are "separable." If they are quicker or more accurate in the correlated condition than in the constant condition, and quicker or more accurate in the constant condition than in the orthogonal condition, then one concludes that the dimensions are "integral." Such a technique lends itself well to the investigation of the recognition of facial identity and facial emotions. If independent mechanisms exist for the processing of these two sorts of facial information, then one should be able to attend selectively to the identity of a face (who the person is) despite the transformations in appearance that that face may undergo as it changes expressions, and be able to attend selectively to the emotion conveyed in a face despite differences in the exact rendering of that emotional expression in different individuals. In Garner's language, if the two functions are truly dissociable, the recognition of facial identity and of facial emotion should be separable for normal perceivers. If the centers mediating these two functions are localized in the right cerebral hemisphere, then damage to this hemisphere should lead to integrality, that is, an inability to attend selectively to facial identity or facial emotion or both (depending on the size and location of the lesion).

To test this idea, photographs of women with happy and sad facial expressions were presented to subjects with right or left hemisphere lesions and to two groups of control subjects. Two decks of cards were constructed: one had photographs of two physically similar women, each displaying either a happy expression or a sad expression; the other had red and blue circles and squares.

Three groups of 12 right-handed subjects were tested using this paradigm: subjects with unilateral right cerebral hemisphere damage; subjects with unilateral left cerebral hemisphere damage; and control subjects who were matched in age, sex, and education with the brain damaged subjects and who had been treated at the same hospitals, but who did not have any cerebral damage. The right- and the left-hemisphere damaged subjects were matched for etiology and locus of lesion (mainly temporal and/or parietal), presence of visual field defects, and paresis or paralysis of the contralateral limbs. Finally, 24 college undergraduates served as a second control group (Table 1).

All subjects were first shown examples of the stimuli and tested for their ability to discriminate them. Then, on each trial, the subject was instructed to sort a deck of 32 cards into two piles.
On different trials, the two piles corresponded either to the two expressions or the two women. In a control task, subjects sorted the geometric figures by color or shape. Each of these types of sorts was done equally often with decks of cards whose dimensions were correlated, orthogonal, or with one

Table 1. Profile of Subjects

Group	Sex	Age	Clinical Picture	Etiology	Localization
RH					
1	M	52	neglect, L hemiparesis, L hemisensory disturb.	CVA(2)	R medial parietal R parasagital frontal
2	M	36	neglect, L hemiparesis, L hemianesthesia	Hemorragic cerebral contusion, MCA CVA	R frontoparietal
3	M	34	L hemiparesis, L hemianopsia	skull fracture MCA CVA	R frontotemporal
4	M	54	L hemiparesis (post resection), L superior quadranopsia	neoplasm	R midposterior temporal
5	M	49	neglect, mild L hemiparesis, L hemianopsia	neoplasm	R posterior temporal R parieto-occipital
6	M	57	neglect, L hemiparesis, L hemisensory disturb	CVA	R deep anterior temporal and parietal
7	F	49	neglect, L hemiparesis, L hemianopsia	CVA	R anterior parietal some extension to frontal
8	M	54	L hemisensory disturb L superior quadranopsia	CVA	R MCA temporal and parietal
9	F	66	neglect, mild L hemiparesis, L hemisensory disturb, L superior quadranopsia	CVA	R MCA temporal and parietal
10	M	61	neglect, mild L hemiparesis, L hemianopsia, L hemisensory disturb	CVA	R MCA temporal and parietal
11	F	67	neglect, L hemianopsia, L hemisensory disturb	CVA	R MCA temporal and parietal
12	M	59	L superior quadranopsia	CVA	R posterior inferotemporal and occipital

Group	Sex	Age	Clinical Picture	Etiology	Localization
LH					
1	M	61	conduction aphasia, R hemiparesis, R hemisensory disturb	CVA	L parietal with some extension to frontal
2	M	23	anomic aphasia, mild L hemiparesis	skull fracture, hematoma evacuation	L temporal, parietal
3	M	62	mild alexia and agraphia R hemianopsia, dyscalculia	CVA	L parieto-occipital
4	M	72	resolving Wernicke's aphasia, R hemiparesis, R hemianopsia	CVA	L MCA temporal and parietal
5	M	58	conduction aphasia, mild R hemiparesis (upper body)	CVA	L frontal insular
6	M	47	resolving Wernicke's aphasia, R superior quadranopsia	neoplasm	L mid posterior temporal
7	F	71	resolving global aphasia R hemiplegia, R hemianopsia, R hemisensory disturb	CVA	L temporal parietal
8	M	54	dyslexia without disgrafia, R hemianopsia	CVA	L parieto-occipital
9	M	48	anomic aphasia, resolving R hemiparesis	CVA	L posterior frontal-temporal
10	M	64	anomic aphasia, R hemiparesis, R quadranopsia, R hemisensory disturb	CVA	L MCA temporal and parietal
11	M	43	anomic aphasia with severe alexia and agraphia, R hemianopsia, R hemiparesis	hematoma AV malformation	L parietal
12	M	52	Wernicke's aphasia, R superior quadranopsia	neoplasm	L temporal parietal

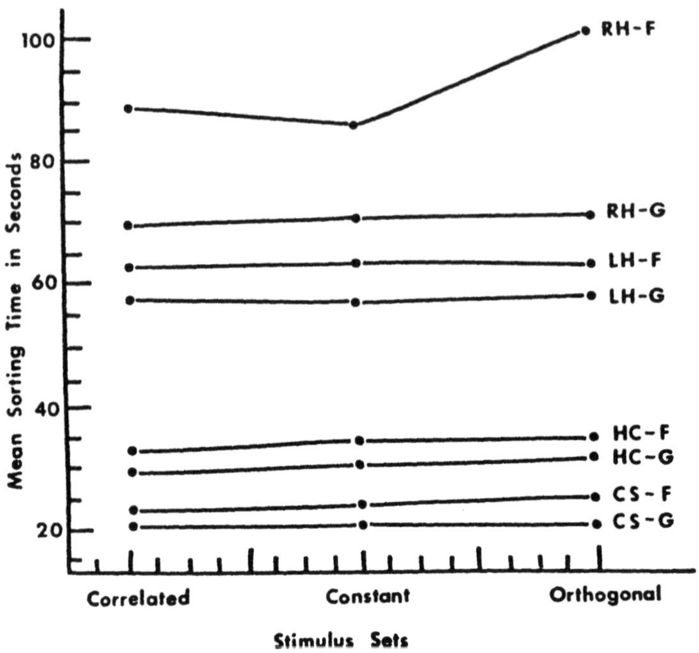

Figure 1. Mean sorting times (sec) for faces and geometric figures

dimension constant. The geometric figures were used to ensure that any differences found between the pattern of results shown by right-hemisphere damaged patients and the other groups was not due simply to deficits in selective attention in general.

The results of this study strongly suggest that the ability to recognize facial expressions of emotion is dissociable from the ability to recognize faces, and that both abilities are mediated largely by the right cerebral hemisphere. First, the right-hemisphere damaged subjects made significantly more errors than did subjects in the other groups both in discriminating identity and in discriminating emotion in faces. Interestingly, the most common error made by right-hemisphere damaged subjects in recognizing identity was to initially see the two women as three or four different people - that is, they had difficulty seeing at least one of the women as the same person when her expression changed.

Second, the left-hemisphere damaged and control groups showed no significant differences among the correlated, constant, and orthogonal conditions when sorting faces. In other words, for these groups, emotion and identity are separable dimensions.
Right-hemisphere damaged subjects, on the other hand, showed orthogonal interference for both types of facial discriminations.
In contrast, the color and shape of the simple geometric figures were separable dimensions for all groups. Therefore, differences between the right

Table 2. Mean sorting times (sec.) for faces: 4 "Fast" RH subjects and 4 "Slow" LH subjects

Group	Dimension	Stimulus Sets: Correlated	Constant	Orthogonal	Mean
"Fast" RH (n=4)					
	Identity	52.8	50.3	54.9	52.7
	Emotion	52.0	53.5	57.8	54.4
	Mean	52.4	51.9	56.4	53.6
"Slow" LH (n=4)					
	Identity	105.6	103.0	102.1	103.6
	Emotion	102.9	105.8	103.9	104.2
	Mean	104.3	104.4	103.0	103.9

hemisphere damaged group and the other groups cannot be attributed to differences in selective attention per se (Fig. 1).

The right-hemisphere damaged group, however, was the slowest of all groups. If the discrimination of identity and of emotion in faces were as difficult for the other groups, would the orthogonal interference seen in the right-hemisphere damaged group when sorting faces emerge for the other groups as well? A careful examination of the data suggests that this would not be the case. First, although the right-hemisphere damaged group was very slow in sorting the geometric figures, it was able to attend selectively to the attributes of these stimuli. Second, when subgroups of the four fastest right-hemisphere damaged and the four slowest left-hemisphere damaged subjects were analyzed separately, the left-damaged subjects were still separable for the faces task and the right-damaged subjects still showed the same pattern of interference. This held true despite the fact that the left-hemisphere damaged subgroup had mean sorting times of about 100 s, well above the mean sorting times of the right-hemisphere damaged group as a whole, and the right-hemisphere damaged subgroup had mean sorting times of about 50 s, well within the range of the left-hemisphere damaged group as a whole (Table 2). Finally, if separate mechanisms within the right hemisphere mediate the abilities to attend selectively to identity and to emotion in faces, then one might expect that different subjects within the right-hemisphere damaged group may have one of these mechanisms damaged and the other intact. If so, one would expect some patients to show integrality only for identity or only for emotion, depending on the size and location of their lesion. Four of the twelve subjects in fact showed such a dissociation. Furthermore, the patients who were integral only for identity had different lesion sites than the patients who were integral only for emotion. The two subjects who were integral only when sorting facial identity both had damage to the posterior right temporal lobe; whereas of the two subjects who were integral only for emotion, one had damage to the right anterior parietal region with some extension to the frontal lobe, and the other had damage to both right temporal and parietal regions.

In sum, this research provides strong support for the notion that the recognition of facial expressions of emotion is dissociable from the recognition of faces per se. Such a finding is consistent with the idea of emotional functions being modular and relatively autonomous from more general cognitive abilities.

It is not altogether surprising that a species dependent on social interaction for survival might have independent mechanisms for the recognition of facial identity and facial emotion, mechanisms which allow the individual to disentangle information relevant to each.

Skill in recognizing faces involves not only the ability to distinguish individuals with highly similar faces, but the ability to recognize a person as being the same individual despite changes in the appearance of the face brought about by age, expression, and angle of view, skin blemishes, or hairstyle. Whatever mechanisms allow us to recognize faces as efficiently as we do must be able to abstract the invariant and distinctive information in a face from that which is transient and nondistinctive.

Similarly, skill in recognizing facial expressions of emotion involves the ability to recognize that expression despite differences in the exact rendering of that expression across individuals. Thus, when recognizing a facial expression on someone we have never met, we must be able to distinguish what is distress from what is a naturally pouty shape of the lips, or anger from a naturally thin straight mouth. A general perceptual device which encoded and attended to both sorts of information would thus be far less efficient than more specialized devices.

Although the finding of a right hemisphere advantage for the recognition of facial expressions of emotion is a robust one, there is some debate as to whether this advantage extends to all facial expressions or just to expressions of negative emotions. This debate provided the context for the next study.

Laterality and Valence of Emotion

Reuter-Lorenz and Davidson (1981) have reported that when happy or sad faces are simultaneously presented with emotionless faces to normal adults, one appearing in each visual field, the happy faces are recognized more quickly by the left hemisphere and the sad faces are recognized more quickly by the right hemisphere. These authors suggest that only negative emotions are more efficiently recognized by the right hemisphere, with positive emotions more efficiently recognized by the left hemisphere.

Although this lateralization by valence (left better at perceiving the positive, right better at perceiving the negative) has not been upheld in the overwhelming majority of studies in which recognition of emotion is the subject, the hypothesis commands attention because it is consistent with some of the literature on asymmetries in emotional mood or behavior following cerebral insult. Studies of mood changes when a single hemisphere has been deactivated by sodium amytal (Perria et al. 1961; Terzian 1964), and of emotional behavior accompanying seizures, lesions, or removal of one hemisphere (Goldstein 1939; Denny-Brown et al. 1952; Hecaen 1962; Gainotti 1972; Sackheim et al. 1982) have generally shown that laughing and/or indifference accompanies the intact left hemisphere, and crying and/or a depressed-worried mood accompanies the intact right hemisphere. These differences are not always found and are open to numerous interpretations (e.g., Milner 1967; Tsunoda and Oka 1976; Geschwind 1981). However, even if it is true that each hemisphere mediates only positive moods and behavior or only negative moods and behavior, the evidence on whether such hemispheric differences exist for perception of emotion as well is equivocal at best.

One way to determine whether the right hemisphere is superior at perceiving facial expressions of emotion or only facial expressions of negative emotions is to have patients with lesions confined to a single hemisphere rate the similarity of facial expressions of positive and negative emotions, and then to submit these ratings to multidimensional scaling analysis (Etcoff 1983, 1984b).

The reasoning behind such a study is this. One can think of the representation of a set of interrelated concepts in terms of a spatial model consisting of an array of points. Each concept (in this case, each emotion) corresponds to a point in space, and the similarity among concepts corresponds to the distance between points. The structure of the entire field of concepts can then be represented by the geometric organization of the points (Shepard 1962a, 1962b). Multidimensional scaling takes as input similarity ratings of pairs of stimuli, and yields as output a "solution" consisting of the coordinates of each stimulus in the psychological "space." Using such spatial models, researchers have found that people's perceptions of facial expressions of emotion are consistently organized along two dimensions: pleasantness, and level of arousal. In the solutions the points representing different emotions roughly form a circle (see Fillenbaum and Rapoport 1971; Russell, 1980; Ekman et al. 1982 for reviews).

If the right hemisphere normally mediates the perception of facial emotions, then right-hemisphere damaged patients should show generally disorganized "emotion spaces" when judging the similarity of the emotions conveyed by such expressions - a random cloud or swarm of points. However, if the right hemisphere is normally involved only in distinguishing among the negative emotions, then right-hemisphere damaged patients should show disorganization only when judging the similarity of negative expressions. Their solutions should look roughly triangular in shape, with a spread of points at the positive pole of the pleasantness dimension (reflecting a preserved ability to distinguish among these emotions), and a tight cluster of points at the negative pole of the pleasantness dimension (reflecting an inability to distinguish among these emotional expressions). Similarly, if left-hemisphere damaged patients are unable to perceive positive facial expressions, one would expect a solution in which the negative pole of the pleasantness dimension is preserved, but the positive pole is collapsed or in disarray.

The same three groups composed of 12 right-handed subjects were tested using this paradigm as were tested in Experiment 1: subjects with unilateral right cerebral hemisphere damage; subjects with unilateral left cerebral hemisphere damage; and control subjects who were matched in age, sex, and education with the right and left hemisphere damaged subjects; and a second control group consisting of college students.

Subjects were presented with two sets of stimulus cards: one had photographs of a woman's face displaying one of nine emotions (happiness, surprise, interest, contentment, fear, anger, sadness, disgust, or neutrality), the other had words printed on it corresponding to the emotions displayed in the faces. All of the photographs were taken from the series used by Ekman and Friesen (1975a, 1975b) who have shown that people identify the emotions in the photographs reliably. I must explain that the "interest" and "contentment" expressions, which are not in the standard Ekman and Friesen set, were included in order to provide further exemplars of positive or pleasant emotional states beyond happiness and surprise (in the standard Ekman series, negative emotions outnumber the positive). I thank Paul Ekman for providing them. In one picture, the model displays a mildly happy expression (this is the contentment expression), and in the other, a blend of happiness and surprise (this is the interest expression). These two pictures were shown to 20 undergraduates who chose "content" as the best descriptor of the mildly Happy expression and "interest" as the best descriptor for the happy/surprise blend.

Subjects were shown all possible pairs drawn from the set of faces, and in a separate part of the procedure, all possible pairs of the corresponding words, and were required to judge the similarity of the emotional states conveyed in each pair. Subjects were then shown each face individually and asked to identify the emotion conveyed. Finally, subjects were questioned about the strategies they used.

Often in multidimensional scaling studies, it is necessary to gather additional data to help interpret the scaling solutions that emerge. To aid later interpretation in this study, the stimuli were rated by two new groups of normal subjects on the following scales: pleasant/unpleasant, aroused/unaroused, dominant/submissive, in control/not in control, intense expression/not intense expression, and attention/rejection. The latter two scales were used in rating faces only. The attention/rejection dimension was defined to the raters largely in terms of the physical appearance of the expression. That is, attention was defined as the active seeking of information with widening of the eyes, nostrils, and mouth, and rejection as the shutting out of information through constriction of the eyes, nostrils, and mouth. Interpretation of dimensions in the scaling solution was then guided by regression of these ratings onto the coordinates of the solutions (see Kruskal and Wish 1978).

The results of this investigation strongly supported the notion that facial expressions of emotion are better recognized by the right hemisphere, regardless of emotional valence. First, once again the right-hemisphere damaged subjects made significantly more errors in identifying emotions than did the other two groups. However, their overall pattern of errors resembled that of the other groups. For all, fear was the most difficult emotional expression to identify, happiness and contentment were the easiest, and fear and anger were the most commonly confused emotions. Right-hemisphere damaged patients made the majority of their errors in identifying negative emotions but so did all subjects. In addition, right-hemisphere damaged subjects were actually slightly better than left-hemisphere damaged patients at recognizing one of the negative emotions - anger. Left-hemisphere damaged subjects did not show a deficit in recognizing positive expressions, as would be predicted by the valence-laterality theory.

Second, the right-hemisphere damaged patients did not show collapsing at either the positive or negative ends of their scaling solutions. Like the other groups of subjects, pleasantness, arousal, and attention/rejection were the perceptually salient dimensions underlying their similarity judgements. The only systematic difference between the right-hemisphere damaged group and the other subject groups, was that their pleasantness dimension was almost entirely redundant with their attention dimension. That is, for the other subject groups the pleasantness dimension was the most salient. This was not true for right-hemisphere damaged patients who conflated pleasantness and attention. Since the attention/rejection dimension was defined to the raters in terms of the physical appearance of the expression, this suggests that perhaps the literal appearance of an expression rather than its underlying significance as a signal of an emotional state may have been most salient to the right-hemisphere damaged group.

This hypothesis received support from an analysis of the strategies reported by the right-hemisphere damaged subjects. Over 75% of these subjects reported that they made attempts to match the physical features of the expressions - the furrow in the brow, the visibility of teeth, whether the mouth was open, etc., in making their similarity judgments. This physical matching was seldom systematic. For example, if the subject found a single feature in common such as teeth being visible, he or she would rate the emotions expressed as similar, even if they were shown in different expressions such as fear and happiness. This strategy was never used by subjects in other groups, and strongly suggests the crucial importance of the right hemisphere for inferring emotional meaning from the pattern of movement of facial muscles. Without an intact right hemisphere, subjects were left analyzing parts of expressions and forced to rely on physical similarity rather than similarity in underlying emotion.

Other than scanning for isolated physical landmarks on the face, the right-hemisphere damaged patients reported trying to recall and imagine situations where particular expressions are commonly seen, and then extrapolating what the emotion must be. The right-hemisphere damaged subjects' level of performance thus depended on several things: the extent to which they scanned the face for the appropriate landmarks (many of which in isolation may mislead the subject in attributing a particular emotion), their knowledge of the correlation between particular landmarks and particular antecedents to emotion, the extent of their knowledge of situations which generate particular emotions, the extent to which they could link the antecedent and the correct emotion, and, finally, the extent to which they thought to use any of the above.

None of these strategies were used by the other groups who relied on direct comparisons of the pleasantness of the emotions to make their judgments.

The left hemisphere damaged subjects did not differ from controls on any of the analyses. In particular, they did not show any selective deficits in the recognition of positive emotional states. If anything, it was the similarity solutions of the right-hemisphere damaged subjects that were more aberrant at the positive than the negative pole.

Interestingly, the results of this study also suggest that the difficulties that right-hemisphere damaged patients have in recognizing facial expressions are dissociable from the ability to conceptualize emotional states, that is, to think about them or reason about them (as opposed to identifying them from visual input). Right-hemisphere damaged patients did not differ from the other subjects in their scaling solutions for emotion words, nor in the strategies they described for judging the similarity of the emotions conveyed in words. Like other subjects, they attempted to assess the relative pleasantness of the emotions and/or the likelihood that the two emotions they were asked to compare would co-occur. In general, all groups judged emotions more similar if they thought they could occur at the same time or easily follow one another in time. This finding suggests that perception of facial expressions of emotion is dissociable from the ability to conceptualize emotional states.

Table 3. Error data

EMOTION	LEFT HEMISPHERE	RIGHT HEMISPHERE
Happy	0	0
Content	0	1
Interest	3	7
Surprise	2	2
Neutral	0	1
Sad	0	4
Angry	2	1
Afraid	4	8
Disgust	1	3

Finally, scrutiny of the error data leaves us with a bit of a puzzle. As one can see, although right-hemisphere patients made more errors in general than did the other groups they were a little better at recognizing one emotion - anger. Why this is true is unclear, and the subject deserves further study. The finding is, however, consistent with Pizzamiglio et al (1983) finding with normal field independent adults that anger may not be better perceived by the right hemisphere (Table 3).

At the very least, the results suggest the following. Most emotions appear to be better perceived by the right hemisphere. The valence of an emotion is clearly an important way in which people organize their perception of emotions. However, the research on perception of facial expression does not provide evidence of opposite lateralization of emotions of different valence. If anything, the results suggest the importance of emotion categories rather than dimensions, with the possibility that certain emotions, such as anger, may have distinct representations from those of other emotions.

Perception of Emotion and Emotional Indifference

Finally, let us turn to the third question, that of the relation among various emotional functions. Several authors have observed profound disturbances in the affective behavior of their right hemisphere damaged patients, in particular a flattening or paucity of emotional behavior. For example, Gainotti (1972) observed unilaterally brain damaged adults during neuropsychological testing. Right-hemisphere damaged patients were rated more often as having "indifference reactions" characterized by denial or minimization of their symptoms and jocularity. In contrast, left hemisphere damaged patients were rated more often as having "catastrophic reactions" characterized by tears, swearing, and expressions of fear and anger. Other emotional behaviors, such as the presence of complex depressive reactions or a tendency to confabulate or harbor delusions, occurred equally often in the two groups.

Several authors have suggested that there may be a relation between the deficits in perception of emotion in patients with right hemisphere lesions and their emotional indifference. For example, both could simply reflect a global disturbance in affect which could accompany right hemisphere damage. Others have suggested a more specific causal link. That is, if right hemisphere damaged subjects are reading the wrong emotion from the face of the people they interact with, then the reactions of the right hemisphere damaged patients may seem inappropriate. For example, some researchers have noted informally that most of the right hemisphere patients they studied who showed deficits in perceiving emotion in faces and in tone of voice also had indifference reactions (e.g., Benowitz et al. 1983). This hypothesis is analogous to hypotheses which have been suggested to account for changes in emotional behavior (such as anger, paranoia, and withdrawal) following hearing loss or impairment (e.g., Zimbardo, et al. 1981).

At the very least, this hypothesis predicts that difficulty in perceiving emotion in faces and indifferent emotional behavior occur in the same right-hemisphere damaged subjects, a correlation that has not been adequately documented in the literature. To test for this correlation, hospital staff members who had worked closely with the patients who had participated in the previously described studies were asked to complete a questionnaire (Etcoff 1983). The questionnaire consisted of statements pertaining to indifference reactions, catastrophic reactions, depressive reactions, and other reactions (including bodily delusions, confusion, or confabulations). In addition, items were included to ensure that raters were responding frankly (i.e., a "lie" scale). These items were taken from Bear and Fedio's (1977) behavioral questionnaire for temporal lobe epileptics.

The results were then analyzed in the following way. First, indices of relative impairment in the perception of emotion in faces were compiled from the results of the two previous experiments on perception of facial expression. These measures, along with a variable for the damaged hemisphere, served as independent variables in a series of regression analyses whose dependent variables were the ratings from the questionnaire. Multiple regressions were performed to determine whether an index of emotion perception predicted variation in emotional behavior above and beyond the variation attributable to hemisphere per se. That is, if the variable corresponding to perceptual difficulty is significant in a regression which includes a variable for which hemisphere was damaged, then one could be more confident that deficits in perception are causally linked to deficits in behavior, as opposed to right hemisphere damage independently affecting perception and affecting behavior. In addition to these analyses, simple correlations between each independent and each dependent variable were computed.

Only one analysis involving either of the two behavioral reactions of interest, the indifference and the catastrophic reactions, reached significance: the simple correlation between hemisphere and indifference, with right-hemisphere damaged subjects more indifferent than left-hemisphere damaged subjects. None of the correlations between degree of impairment in perceiving emotions and degree of indifference were significant, nor was the correlation between hemisphere and ratings on the catastrophic reaction.

The results replicated the findings of Gainotti (1972) and others that indifference is more common among right-hemisphere damaged subjects than among left-hemisphere damaged subjects. In addition, it provided evidence that indifference and difficulty in perceiving emotions in the face can occur in the same right-hemisphere damaged patients, but also that the occurrence of one does not necessarily imply the presence of the other. This finding suggests that production of complex emotional behaviors and the perception of facial expressions of emotion are mediated by separate circuits within the right hemisphere. In cases where both circuits are damaged, difficulties in perceiving emotions may exacerbate the behavior problems experienced by these patients, but it is unlikely that problems in perceiving emotion cause such behavior problems.

Before attempting to summarize the findings, I would like to make one brief point. These studies have all employed posed pictures of facial emotion, and thus do not address the question of recognizing voluntary vs involuntary emotional displays. I would like to hazard the guess, however, that the results presented here could be even stronger for involuntary expressions. Paul Ekman and others have argued that lateralization of emotional expression should increase with more voluntary control of the expression (e.g. Ekman 1980; Ekman et al. 1981). However, on the recognition end, quite the opposite may be true for the following reason. A device which would be truly useful, in other words good at recognizing emotion, would have to be one which could discern emotion even in cases where the other person was trying to hide that emotion, so that the emotional display provided false information about intentions. If our emotion detectors could not uncover the anguish behind the brave front or discern true delight from a polite smile, it would be a poor device indeed.

IN SUMMARY

I will now try to summarize the findings from these three studies. First, the perception of facial expression of emotion appears to be mediated largely by the right cerebral hemisphere, and appears to be largely independent of cognitive skills believed to be subserved by the right hemisphere, such as the recognition of the identity of faces and of simple geometric patterns. Second, at least for the recognition of facial expression, valence does not seem to be a "psychologically real" entity which can be localized to one hemisphere of the brain. If anything, the results of study two provide more evidence for the localization of emotion categories (such as anger vs sadness) rather than dimensions. Third, the intact right hemisphere appears to mediate the recognition of facial expressions of emotion by analyzing the expressions in terms of the underlying state that complex combinations of features convey rather than by the physical properties of the features themselves, and by its knowledge of what is distinctive to each expression. Fourth, the ability to recognize facial expressions of emotion communication appears to be relatively independent from other emotional abilities. It is not correlated with the ability to show proper emotional response and concern about one's own well-being nor to the ability to understand the meaning of emotion terms. It is probably thus not terribly useful to think of "global" deficits in emotion and to expect that findings in one arena (for example, emotional behavior) should hold for other arenas as well (for example, emotion recognition).

Getting back to our original questions, the findings presented here are consistent with models of emotion which stress its autonomy.
The findings also provide evidence that at least some emotion functions are lateralized - that is, show right-hemisphere superiority. Finally, they suggest that emotion may best be thought of as a relatively autonomous system which itself is composed of relatively autonomous subsystems or modules.

Acknowledgements. Preparation of this paper was supported by NIH grant 1RO1HD22166, and by the MIT Center for Cognitive Science under a grant from the Alfred P. Sloan Foundation. I thank Guido Gainotti, the II University of Rome, and the Istituto della Enciclopedia Italiana for their invitation to participate in this Conference emotional expression in others correlates with the ability to show appropriate emotional concern about one's own well being.

REFERENCES

Bear DM, Fedio P (1977) Quantitative analysis of interictal behavior in temporal lobe epilepsy. Arch Neurol 34: 454-467

Benowitz LI, Bear DM, Mesulam MM, Rosenthal R, Zaidel E, Sperry, RW (1983) Nonverbal sensitivity following lateralized cerebral injury. Cortex 19: 5-12

Bruyer R, Laterre C, Seron X, Feyereisen P, Strypstein E, Pierrard E, Rectem, D (1983) A case of prosopagnosia with some preserved covert remembrance of familiar faces. Brain Cogn 2: 257-284

Cicone M, Wapner W, Gardner H (1980) Sensitivity to emotional expressions and situations in organic patients. Cortex 16: 145-158

Darwin C (1872) The expression of the emotions in man and animals. John Murray, London. (Current ed, University of Chicago Press, Chicago 1965)

Dekosky ST, Heilman KM, Bowers D, Valenstein E. (1980) Recognition and discrimination of emotional faces and pictures. Brain Lang 9: 206-214

Denny-Brown D, Meyer JS, Horenstein S (1952) The significance of perceptual rivalry resulting from parietal lesions. Brain 75: 433-471

De Renzi E, Spinnler H (1966) Facial recognition in brain-damaged patients: an experimental approach. Neurology 16: 145-152

Ekman P (1980) Asymmetry in facial expression. Science 209: 833-834

Ekman P, Friesen WV (1971) Constants across cultures in the face and emotion. J Pers Soc Psychol 17: 124-129
Ekman P, Friesen WV (1975a) Pictures of facial affect. Consulting Psychologists Press, Palo Alto, California
Ekman P, Friesen WV (1975b) Unmasking the face. Prentice Hall, Englewood Cliffs, NJ
Ekman P, Hager J, Friesen WV (1981) The symmetry of emotional and deliberate facial action. Psychophysiology, 18: 101-106
Ekman P, Friesen WV, Ellsworth P (1982) What emotion categories or dimensions can observers judge from facial behavior? In: Ekman P. (ed.) Emotion in the human face. (2nd ed) Cambridge University Press, Cambridge
Etcoff NL (1983) Hemispheric differences in the perception of emotion in faces. Unpublished doctoral dissertation, Boston University
Etcoff NL (1984a) Selective attention to facial identity and facial emotion. Neuropsychologia 22: 281-295
Etcoff NL (1984b) Perceptual and conceptual organizaton of facial emotions: hemispheric differences. Brain Cogn 3: 385-412
Fillenbaum S, Rapoport A (1971) Structures in the subjective lexicon. Academic, New York
Fried I, Mateer C, Ojemann G, Wohns R, Fedio P (1982) Organization of visuospatial funtions in human cortex. Brain 105: 349-371
Gainotti G (1972) Emotional behavior and hemispheric side of lesion. Cortex 8: 41-55
Garner WR (1974) The processing of information and structure. Erlbaum, Hillsdale
Garner WR (1976) Interaction of stimulus dimensions in concept and choice processes. Cogn Psychol 8: 98-123
Gazzaniga MS, Smylie, CS (1983) Facial recognition and brain asymmetries: clues to underlying mechanisms. Ann Neurol 13: 536-540
Geschwind N. (1981) The significance of lateralization in nonhuman species. The Behav Brain Sci 4: 26-27
Goldstein K (1939) The organism: a holistic approach to biology, derived from pathological data in man. American Books, New York
Hansch EC, Pirozzolo FJ (1980) Task relevant effects on the assessment of cerebral specialization for facial emotion. Brain Lang 10: 51-59
Hecaen H (1962) Clinical symptomatology in right and left hemisphere lesions. In: Mountcastle VB (eds) Interhemispheric relations and cerebral dominance. Johns Hopkins, Baltimore, Md
Kruskal JB, Wish M (1978) Multidimensional scaling. Sage, Beverly Hills, California
Landis E, Assal G, Perret E (1979) Opposite cerebral hemispheric superiorities for visual associative processing of emotional facial expressions and objects. Nature 278: 739-740
Ley RG, Bryden MP (1979) Hemispheric difference in processing emotions and faces. Brain Lang 7: 127-138
Milner, B (1967) Brain mechanisms suggested by studies of the temporal lobes. In: Milikan CH, Darley FL (eds) Brain mechanisms underlying speech and language. Grune & Stratton, New York

Perria L, Rosadini G, Rossi GF (1961) Determination of side of cerebral dominance with Amobarbitol. Arch Neurol 4: 173-181
Pizzamiglio L, Zoccolotti P, Mammucari A, Cesaroni, R (1983) The independence of face identity and facial expression recognition mechanisms: relation to sex and cognitive style. Brain Cogn 2: 176-188
Reuter-Lorenz P, Davidson RJ (1981) Differential contribution of the two cerebral hemispheres to the perception of happy and sad faces. Neuropsychologia 19: 609-613
Rizzolatti G, Umilta C, Berlucchi G (1971) Opposite superiorities of the right and left cerebral hemispheres in discriminating reaction time to physiognomical and alphabetical material. Brain 94: 431-442
Russell JS (1980) A circumplex model of affect. J Pers Soc Psychol 39: 1161-1178
Sackheim HA, Greenberg MS, Weiman AL, Gur RC, Hungerbuhler JP, Geschwind, N (1982) Hemispheric asymmetry in the expression of positive and negative emotions. Arch Neurol 39: 210-218
Shepard RN (1962a) The analysis of proximities: multidimensional scaling with an unknown distance function: I Psychometrika 27: 125-140
Shepard RN (1962b) The analysis of proximities: multidimensional scaling with an unknown distance function: II. Psychometrika 27: 219-246
Shuttleworth EL, Syring V, Allen N (1982) Further observations on the nature of prosopagnosia. Brain Cogn 1: 307-332
Strauss S, Moscovitch M (1981) Perception of facial expressions. Brain Lang 13: 308-332
Terzian H (1964) Behavioral and EEG effects of intracarotid sodium amytal injections. Acta Neurochirugie 12: 230-239.
Tsunoda T, Oka M (1976) Lateralization for emotion in the human brain and auditory cerebral dominance. Proc Jap Acad 52: 528-531
Warrington, EK, James M (1967) Disorders of visual perception in patients with localized cerebral lesions. Neuropsychologia 6: 253-266
Zimbardo PG, Andersen SM, Kabat LG (1981) Induced hearing deficit generates experimental paranoia. Science 212: 1529-1531

Disturbance of Facial Emotional Expressions in Brain-Damaged Subjects

Luigi Pizzamiglio and Alessandra Mammucari

Department of Psychology, University of Rome La Sapienza, Rome, Italy

INTRODUCTION

The neuropsychological approach to the study of emotion is justified on two grounds: first it is of primary importance in gaining further insight into the nervous mechanisms underlying the organism's capacity for lively and immediate response to rapid and complex changes in the environment; second, the observation that adaptive capacities are affected by brain damage makes it possible to gather evidence which can be used to refute or confirm the numerous psychobiological theories regarding the interpretation of emotional behavior. Numerous observations of the asymmetry with which the two brain hemispheres seem to interfere in the organization of emotional responses have led neuropsychologists to study the visible consequences in the presence of focal brain damage or after drug manipulation in various brain territories (Etkoff 1986; Gainotti, this volume). This approach may be related back to the first area of neuropsychological interest. The results obtained so far have repeatedly pointed to a predominant role of the right hemisphere in emotional control or, alternatively, a differential involvement of both hemispheres in the response to different emotional stimuli. At the present state of our knowledge, however, the sum total of available evidence does not provide a universally interpretable picture with regard to the regulation of the motor response and autonomic mechanisms of the individual and to the mechanisms for decoding external events to which the individual must respond.

One of the objectives of the present work is to make as detailed a study as possible of the contribution made by each hemisphere, or of more circumscribed structures inside each hemisphere, to regulating emotional and nonemotional responses of the human face. The specific attention accorded to the emotional expressiveness of the human face was historically evoked by Darwin and is justified by the extraordinary importance and accuracy with which individuals signal their emotional reactions in their intercourse with other members of their species.

The issue of the neural control of facial expression can be further specified in a number of different questions. The first one concerns the existence of common or separate control for facial movements in general and for a special

Experimental Brain Research Series 18

category of actions involved in mimicry. These latter actions may be used for social communication or, specifically, to convey emotional information; each of these actions may be controlled by the same or by different mechanisms. Second, it is well extablished, from a variety of clinical observations, that voluntary facial expression can be disrupted independently from spontaneous expression of emotions (Rinn, 1984). The question is whether hemispheric specialization applies to both these classes of behaviors.
An additional, although less explored, set of questions refers to the relations between the control of facial emotional expressiveness and other manifestations of emotions, such as the autonomic responses, the decoding of emotionally relevant input and the emotional feeling and moods often induced by a particular experience. Experimental and clinical neuropsychology can contribute toward clarifying these issues which have relevant implications for the construction and validation of theories of emotion in humans.

As for the first issue, a number of studies on normal subjects have looked for the existence of differences between the two halves of the face in performing unilateral or bilateral facial movements. These attempts are based on the assumption that any visible and systematic facial asymmetry can be considered a consequence of a superiority of the controlateral hemisphere in programming facial movements.
A number of studies showed that the left hemiface was judged to be more expressive when subjects were asked to perform unilateral movements, such as winking, and pulling the mouth up and down (Chaurasia and Geswami 1975; Koff et al. 1981; Borod et al. 1983; Alford 1983) or bilateral facial actions (Campbell 1982). Nevertheless, when normal adults were requested to imitate a large number of action units, such as wrinkling the nose, raising the eye brows etc., no systematic bias toward either side of the face was found (Hager and Ekman 1985). The important difference between the two groups of research is that the former scored facial movements using a subjective scale of intensity of expression, whereas the latter used the FACS scoring system, which is a more objective and quantifiable procedure.

The clinical literature does not provide any systematic description of disorders of facial mimicry. Occasionally, together with a description of the well known oral apraxia, there are comments about the difficulty observed in some patients in intentionally producing in imitating mimic facial movements (Natham 1947).
Altogether, the available data do not univocally support the above mentioned assumption of right hemisphere control for non emotional facial movements.
A first goal of the present paper is to look at comparison between left and right focal brain lesions in producing a differential impairment in imitating a large set of facial action units without emotional value. The study will also look for a relationship between imitation of mimic facial actions and the presence of oral apraxia.

The literature on the production of facial emotional expressions produced a greater number of studies which are addressed either to intentional or to spontaneous facial expressions. In studying voluntary emotional expression the technique of making composite pictures of two left or two right hemifaces of the same subjects performing one emotion was extensively used. The presentation of chimeric faces showed that negative emotions were judged to be more intense for the left composite pictures (Sackeim et al. 1978). Similar evidence was also given for smiling, both in adults (Campbell 1978) and in children (Rubin and Rubin 1980), although it was not confirmed by Sackeim et al.'s study (1978).

A left facedness was described by a number of studies, which indicated a greater asymmetry toward the left side in producing voluntary expressions (Borod and Caron 1980; Borod et al. 1981).

Similar to the imitation of nonemotional facial actions, no bias toward either side was found when an objective technique was used to code facial actions (Ekman et al. 1981).
In brain damaged patients, Borod et al. (1985) showed a greater impairment of right lesions, but the same finding was interpreted as a direct consequence of facial paralysis by Bruyer (1981).

Spontaneous facial expressions of emotion produced a left facedness in a number of studies (Moscovitch and Olds 1982; Dopson et al. 1984; Borod et al. 1983; Scwartz et al. 1979), but not in others (Lyn and Lyn 1938; Ekman et al. 1981); the picture varied according to the positive-negative dimension of the emotions.

Evidence from the clinical literature points to a greater right hemisphere impairment in expression of emotions (Ross and Mesulam 1979; Buck and Duffy 1980; Borod et al. 1985). These data were not replicated by Kolb and Milner (1981). A review of the cases on pathological laughing and crying and on gelastic and dacristic epilepsy were considered in line with a greater involvement of the right hemisphere in producing negative and of the left hemisphere in producing positive emotional expression (Sackeim et al. 1982). A direct observation of epileptic outbursts of laughing and crying during the crisis did not confirm the previous finding (Straus et al. 1983).
The available data on facial expression of emotions, either spontaneously or intentionally produced, is not yet univocally supportive of the lateralization of the control for facial expressive behavior.

A second contribution reported in the paper refers to a more systematic study of spontaneous facial expression in response to emotional stimuli in a large population of focal brain lesions to the left and right hemisphere. An objective coding for measuring facial activity was used, in contrast to all other previous research.

The third point raised earlier concerns the limited number of investigations on autonomic responses to emotional stimuli and the scarcity of studies in which different measures of emotional behaviors were intercorrelated. Previous studies pointed out that the autonomic concomitants were clearly reduced in right brain damaged patients when compared with left hemisphere patients and controls (Heilman et al. 1978; Morrow et al. 1981; Zoccolotti et al. 1986).

The same brain-damaged patients used in this paper to study the imitation of facial movements and the production of spontaneous facial expression have also been tested for their autonomic responses (i.e., heart rate and galvanic skin conductance). The report of these findings will be given in a different chapter in this book (see the chapter by Caltagirone and Zoccolotti). For this reason, the following plan of the study serves both to introduce the research reported in the present chapter, as well as the experiment on the relation between autonomic responses and facial behavior described in the following chapter.

General Plan of Study

Three different groups of subjects were examined: one comprising control subjects without brain damage, but comparable to the other groups for age, sex and education (28); one comprising brain-damaged patients with brain lesions located in the left hemisphere (39); and one comprising patients with unilateral right hemisphere lesions (23). All the patients in the last two groups displayed cerebrovascular lesions and were examined at the Neurological Clinic of the Universita' Cattolica of Rome. Two criteria were used for deciding whether or not to include patients in the study: the presence of a single lesion in either hemisphere and the capacity to understand the tasks they were asked to carry out.

Description of the Experimental Set-up

The experimental set-up was designed to allow for the collection of samples of facial expressions produced spontaneously, intentionally and by imitation. At the same time, provision was made for recording autonomic responses produced while the subjects were responding to the stimuli presented to them. Each subject entered a silent room, which was normally used for clinical electrophysiological recordings, and seated in such a way as to hide any videotape-recording apparatus; electrodes for electrocardiograph and skin conductance were applied. The subject was told that an electrocardiograph would be taken under special stimulation conditions and then left alone in the dimly lit room in front of a translucid screen. From outside the room, short film sequences were projected for the purpose of inducing spontaneous emotional responses in the subject; all the facial movements produced by the subject were videotaped. The films used consisted of four sequences (each of less than 1 minute's duration) which had produced homogeneous emotional responses in all the subjects examined in previous experiments.

The first sequence, intended to produce pleasurable sensations in the spectator, was taken from a commercial: a thin man and a fat man are talking together; the thin man runs away and this gives rise to some comic scenes. A second sequence showed the surgical medication of a serious face burn: the extreme crudeness of the scenes aroused strongly negative reactions, such as disgust, fear, stress, in most of the subjects. A third sequence showed images that would normally arouse no emotions, either positive or negative: it shows waves gently breaking on a beach. The fourth sequence, showing a puppy playing with a flower held in the actor's hand, normally aroused a positive and happy sensation.

The four film sequences were preceded and followed by the presentation of a further four film sequences. The addition of the latter stimuli served two purposes: to introduce and accustom the subject to the situation, before the presentation of the critical sequences, and to dispel any influence of one target on the following one.
After the projection of the films, the experimenter entered the room and asked a number of questions aimed at ascertaining whether the subject had understood the meaning of the four critical film sequences. Subjects found not to have understood the meaning of the films clearly were excluded from the experiment.

The subject, now informed that his/her facial movements were being observed, was then shown a short film sequence in which an actor performs several isolated facial movements, for instance, turning up his nose, raising the eyebrows, dropping the corners of the mouth, and so on. The subject had the task of imitating the actions appearing on the screen. The subjects received verbal advice from the experimenter in order to simplify their task (for example, if when raising their eyebrows they tended also to open their eyes, they were told: "Only raise your eyebrows, don't open your eyes").

Throughout the procedures aiming at stimulating the production of spontaneous or intentional facial expressions or the imitation of movements, the heart rates of the subjects examined and the psychogalvanic responses (on the hand homolateral with the brain damage) were also recorded. In a separate setting the same subjects were given a battery of neuropsychological tests including visual spatial ability, face and emotional expression recognition, and three tests of oral, constructive and ideomotor apraxia. In the present chapter the results on the imitation of facial movements and the production of spontaneous emotional responses are presented.

Imitation of Facial Movements ([1])

The general questions raised at the outset also include several more specifically related to the problem of imitating facial movements. The first of these concerns the possibility of facial action being affected more by damage to one hemisphere or to the other. So far no systematic research has been carried out on this subject and many researchers consider that these movements are controlled in more or less the same way as oral movements,

[1] Prepared in collaboration with C. Caltagirone, P. Ekman, and W.V. Friesen.

Table 1. Imitation of facial actions

Upper part of face
Raises inner corner of brow
Raises outer corner of brow
Raises entire brow
Lowers and pulls brow together
Squints eyes, makes crowsfeet wrinkles
Squints eyes, raises and straightens lower lid
Lower part of face
Pulls lower lip down
Common smile
Chin raiser
Lip pressor
Dimples cheeks
Stretches lip corners straight to the side
Raises upper lip
Wrinkles nose

which are regularly examined during the clinical assessment of oral apraxia. The second question bears upon the special relationship between the control of facial action and specific locations inside each hemisphere. Then there is the further question of whether any systematic relationship exists between the imitation of facial actions and other forms of oral apraxia, both constructional and ideomotor.

In the experimental framework outlined above, the subject was asked to imitate fourteen facial actions (Table 1). Evaluation of patients' responses was done using REFACT, as developed by Ekman and Friesen (1978). The latter consists of a four-point scale to evaluate the adequacy of the patients' performance in each of the imitated actions (In this scale, 1 stands for accurate imitation (only the requested facial action is performed); 2 for accurate imitation with the addition of "non-facilitatory" facial actions; 3 for accurate imitation with the addition of "facilitatory" facial actions; and 4 for requested movement not performed). An evaluation is also made of any asymmetry in the performance of facial actions.

All the brain-damaged patients were also subjected to an ideomotor apraxia test (De Renzi et al. 1968), a constructional apraxia test (Gainotti et al. 1977), and an oral apraxia test (De Renzi et al. 1966).

Results

Separate analyses were carried out on the subjects' performances using the scoring system outlined above. When point score 1 is used (i.e., a restrictive

Table 2. Imitation of facial movements: mean values and standard deviations (in parentheses).

	Restrictive scoring (1)	Intermediate scoring (1+2)	Permissive (1+2+3)
Normal subjects	3.17 (2.12)	5.17 (2.29)	13.06 (1.23)
Right brain-damaged	2.77 (2.22)	4.43 (2.12)	11.73 (1.76)
Left brain-damaged	2.75 (1.89)	5.08 (2.17)	12.69 (1.28)
	$F=0.33$, n.s.	$F=0.66$, n.s.	$F=4.86$, $p<0.009$

evaluation is made), the task is found to be particularly difficult by all three subject groups. The overall performance for all groups ranges from 20% to 25% correct answers. Analysis of variance of the total scores obtained in the test shows that the control subjects are slightly but not significantly better at performing this task than either right or left brain-damaged subjects.

The performance of the three groups increases considerably if a more lenient scoring system is used, i.e., if imitations are accepted by movements not facilitating the one requested (point score 1 + 2), the performance of the 3 groups of subjects improves and the number of correct responses rises to about 30%-35%. Comparison between the three groups using analysis of variance reveals no significant differences between control subjects and right or left brain-damaged subjects. The performance of the three groups becomes very high when the most permissive scoring system is used, i.e., all correct imitations are accepted even when accompanied by non-requested movements. In this case the correct performance of the items requested approaches 90%. Analysis of variance indicates a significant group factor effect: the performance of the controls and left brain-damaged subjects is slightly better than that of right brain-damaged ones (Table 2).

The imitation responses for movements of both the upper and lower parts of the face were analysed separately in a subsequent test series. In both cases the use of a restrictive scoring system (1 and 1+2) revealed no difference between the three groups of subjects. On the other hand, the use of the most permissive scoring system (1+2+3) revealed that imitation of movements involving the upper part of the face are performed better by the control group and by the left brain-damaged group than by the right brain-damaged group, while in imitation of movements of the lower part of the face, only the control group differs significantly from the brain-damaged group, whose mean scores coincide almost exactly.
In a further test, the time elapsing between the presentation of the filmed stimulus and the moment in which the patient succeed in performing the required movement correctly was recorded. The mean and standard deviations of the performance times for the three groups are shown in Table 3. Analysis

Table 3. Latent time (in seconds) to obtain best possible imitation of the four movements requested: mean and standard deviations (in parentheses)

Normal subjects	2.54 (0.46)
Right brain-damageds	3.07 (0.88)
Left brain-damageds	2.52 (0.63)
	$F = 7.71$, $p < 0.001$

of variance reveals that the right brain-damaged group is slower than the left brain-damaged and the control groups.
It was also attempted to identify a score below which the brain-damaged patients could be identified as having a specific disturbance in the performance of miming movements. Using point scores 1 and 1+2 it proved impossible to find a score which could be used to discriminate between "normal" and "pathological" performances. However, if the 1+2+3 system was used, no control subjects obtained a score of less than 10 items out of 14; three right brain-damaged subjects and one left brain-damaged subject had scores of less than 10.

It is noteworthy that the difference between right and left brain-damaged groups and controls is quantitatively small, particularly when compared with the large differences between normal and pathological performance found in other forms of apraxia. The patients' performance in imitating facial movements was compared separately with the presence of forms of oral, ideomotor and constructional apraxia.

The substantial lack of correlation between the imitation of facial movements and the other forms of apraxia is clearly shown in Table 4.
In particular, one patient with left brain damage is simultaneously affected by oral apraxia and has a lower cut off score in the imitation of facial movements test. In all the other patients the two tests are always dissociated. Similar observations can be made with reference to ideomotor and constructional apraxia (see Table 4).
CAT scans were available for 33 brain-damaged patients (13 with right lesions, 14 with left lesions and with aphasia, and 6 with left lesions without aphasia).
On the basis of a lateral projection reconstruction of the lesion, the size was assessed as large, medium sized, or small. Its relative distribution in the right and left hemisphere is shown in Table 5.
There is no apparent difference in lesion size in the two groups. The relationship between lesion size and imitation of facial movements also shows that, quite apart from the scoring system used, large and medium-sized lesions tend to be associated with worse performances than the small lesions, but in no case are the contingency coefficients significant.

Table 4. Contingency table for presence or absence of different forms of apraxia and performances in facial movement imitation tests

	Imitation of facial movements (cut-off = 10)			
	cut-off	top	bottom	
Oral apraxia	(9)			
(n = 58)	top	40	3	Chi^2= 0.001
	bottom	14	1	C =0.005
Ideomotor apraxia	(12)			
(n = 36)	top	27	4	Chi^2= 0.75
	bottom	5	0	C =0.14
Constructional apraxia	(66)			
(n = 44)	top	34	4	Chi= 0.69
	bottom	6	0	C =0.12

Table 5. Extension of lesions in brain-damaged subjects

	Small lesion	Medium-sized lesion	Large lesion	
Right brain-damaged	0	8	9	Chi^2=4.25 p=0.12
Left brain-damaged	5	9	9	C =0.31

The role played by lesion location was determined by examining the relevance of lesion presence/absence in a cerebral lobe to the performance of facial movements (for instance, patients with parietal lobe lesions were compared with all the patients with no parietal lobe involvement). Although the number of patients with single lobe involvement is too small for any definitive conclusions to be drawn, none of the tests carried out indicated that any specific location was important in the genesis of this disorder. It can thus be concluded that lesion size may play a marginal role in influencing brain-damaged patients' performance in tasks involving imitation of facial movements, while there is no evidence of specific relevance of cortico-subcortical location to performance in such tasks. Lastly, the variety of locations of the lesions in the four patients with the poorest performances indicated that there are no privileged areas for a given form of facial apraxia.

In summarizing the results illustrated above, several final considerations can be made:

1. Focal lesions in either hemisphere have no effect on the performance in the imitation of isolated facial movements: the performance of the brain-damaged patients is indistinguishable from that of normal subjects as regards the capacity to produce bilateral contractions of individual facial muscles in an imitation task.
The performance of control subjects and of brain-damaged patients improves when movements facilitating the requested action are used. Patients with right cerebral lesions seem to be slightly less efficient in using these facilitatory mechanisms and are slower to produce a suitable facial imitative response.

2. The imitation of facial movements in brain-damaged patients seems to be quite unrelated to their overall praxic capacity. This conclusion is particularly important in view of the double dissociation between poor performance in facial imitation tasks and oral apraxia. In the populations studied there does not seem to be any evidently pathological performance in the imitation of facial movements; this result is in contrast to the great frequency with which oral, ideomotor, and constructional apraxia deficits are encountered in the same brain-damaged population.

3. The facial imitation task does not seem to be selectively disturbed by the presence of cortical and subcortical lesions located in either hemisphere; there is no correlation between the size of the lesion and the disorder itself. This data set seems to suggest that, unlike the case of mouth and limb movements, the control of facial imitation movements may be partially dependent on the pyramidal system, although it may also have a multiple and diffuse representation in the extrapyramidal system, in a similar fashion to Geschwind's hypothesis (1975) concerning both body axial and extrinsic eye movements. This interpretation would lend support to the rare clinical descriptions of facial imitation disorders in brain-damaged patients. The results of this research represent an important contribution to the study of asymmetries in the production of facial expressions, which has alternately been attributed to either hemisphere in the past. The results do not support the notion that the right hemisphere is more deeply involved in the production of all the facial movements (Etkoff 1986), while they agree with the idea of Hager and Ekman (1985) that the asymmetries between the two hemifaces cannot be interpreted in terms of cerebral asymmetry.

Production of Spontaneous Emotional Facial Expressions ([2])

This section sets out the results referring to the production of spontaneous emotional facial expressions in the same group of patients. The general questions raised in the introduction can be specified and restated as follows:

---

[2] Prepared in collaboration with C. Caltagirone, G. Gainotti, P. Ekman, W.V. Friesen, and P. Zoccolotti.

1. Is there a difference between the spontaneous emotional expressions produced in different hemispheres? It has already been noted in the introduction that the three relevant investigations into this question have given mutually inconsistent answers (Buck and Duffy 1980; Kolb and Milner 1981; Borod et al. 1985). Furthermore, all these studies are characterized by the use of small groups of patients, by the subjective evaluation of the facial expressiveness and, in particular, by an evaluation of the emotional involvement of the patients under observation.

2. Does there exist a specific location which produces more frequent disorders of spontaneous emotional facial expressions? Samples of spontaneous facial expressions were collected using the techniques described above. The emotional facial responses were evaluated using the FACS technique developed by Ekman and Friesen (1978), in which the analysis units consist of the smallest identifiable discrete actions produced by the facial muscles. Each of these facial actions is evaluated in terms of intensity (on a point scale of 1 to 5) and in terms of asymmetry.
In order to decide whether a given combination of facial actions can be considered representative of one of six primary emotions (fear, joy, sadness, disgust, surprise, and anger), the scores obtained by the subjects in this test were compared with the "encyclopaedia" of human facial activities developed by Ekman and Friesen. Using the above criteria the emotional expressions were subdivided into the following categories: (a) clearly identifiable emotional patterns produced in response to film stimuli; these patterns of facial action may be accompanied by additional actions without any particular emotional significance or by mixed emotional expressions (a happy face overshadowed by surprise or sadness, etc.); (b) nonemotional movements (for example, slight crooking of the neck or backward and forward movement of the head); and (c) avoidance responses, i.e., movements that can be interpreted intuitively as tending to ward off the film stimulus presented (e.g., the subjects look away, lower their heads, turn a round, or else close their eyes for a fairly long time: 3 s. or more).

In order to check whether an examination of the results obtained using such an analytical method would produce different results to the one carried out using literature methods, the patients' performance were also subjected to two successive analyses using the evaluation procedures described by Buck and Duffy (1980) (subject point scale from 1 to 7) and by Borod et al. (1985) (1-5 point scale).
Analyses of "nonemotional movements" showed a significantly greater facial mobility in the control subjects compared with the two brain-damaged groups. As evaluated by FACS, the facial emotional responses are more frequently elicited by positive film sequences (i.e., happy ones) than by unpleasant or neutral sequences. The responses of the three groups are shown in Table 6. For the two pleasant film sequences no significant differences are found between the three groups of subjects, while a significantly higher number of emotional responses are given by the control group for unpleasant or neutral sequences. No appreciable differences were found between right and left brain-damaged subjects. Using the other two measurement methods based on subjective

Table 6. Number of subjects producing or not producing emotional facial responses in the three groups of subjects
(C = control subjects; R = right brain-damageds; L = left brain-damageds).

Film sequences												
Emotional response	Dog playing			Fat man thin man			Burn			Waves		
	C	L	R	C	L	R	C	L	R	C	L	R
Present	14	16	7	11	18	10	14	7	5	11	4	0
Absent	14	23	16	17	21	13	14	32	18	17	35	23
	$Chi^2 = 1.9$ n.s.			$Chi^2 = 0.31$ n.s.			$Chi^2 = 8.91$ $P<0.01$ C>R C>L R=L			n.s.		

Table 7. Number of subjects producing or not producing avoidance responses during the four film sequences in the three subject groups
(C = control subjects; R = right brain-damaged; L = left brain-damaged).

Film sequences												
Avoidance response	Dog playing			Fat man thin man			Burn			Waves		
	C	R	L	C	R	L	C	R	L	C	R	L
Present	2	3	1	2	1	3	12	3	14	7	4	7
Absent	26	20	38	26	22	36	16	20	25	21	19	32
							$Chi^2 = 5.56$ $P<0.06$ C>R C=L R<L			$Chi^2 = 0.63$ n.s.		

scoring (Buck and Duffy 1980; Borod et al. 1985), no differences were detected in the emotional responses of the three groups of subjects, whether all the films were considered together or separately.

"Avoidance" responses are more numerous in the presence of unpleasant sequences (man with the burnt face) compared with neutral and positive films. In right brain-damaged subjects this behavior is much less frequent than in left brain-damageds and controls (Table 7). This result clearly indicates that

when normal subjects and left brain-damaged subjects receive a very unpleasant stimulus they either produce an appropriate facial expression or behave in such a way as to avoid the stimulus: the first of these behaviors, although not the second, is found also in right brain-damaged subjects. Although clearly showing that they have understood the significance of the stimulus presented to them, the latter do not seem inclined to do anything to eliminate such a disagreeable situation. A further investigation involving the scoring of the negative responses by the three groups of subjects revealed no significant differences between right and left brain-damaged subjects.
As in the preceding investigation of facial movements, the brain-damaged subjects were subdivided according to whether they suffered from frontal, parietal, or occipital lesions. An attempt was made to ascertain whether the existence of a lesion in one of these lobes led to differences in the production of facial emotional expressions. No difference was found between location in different lobes in either hemisphere with regard to the production of emotional responses. Since, in two of the works cited above, worse performance was found in patients with frontal lesions, the latter patients were subjected to a separate study. Four patients had very extensive prefrontal lesions, two had lesions in the left hemisphere and two in the right. In one of the subjects with a left prefrontal lesion and in one subject with a similar right lesion, the emotional responses were completely absent; in the other two cases they were present as normal.

In summing up the results of this second experiment, three main conclusions can be drawn. One important conclusion is negative: quite independently of the method used (subjective assessment or highly analytical evaluations), patients with unilateral right or left lesions showed no difference in the acceptable emotional responses produced; likewise, no differences were found in the intensity of the emotional expressions produced. The results of the present investigation nevertheless seem to indicate that the FACS scoring method is much more sensitive than the subjective methods used in previous research work to capture intersubject differences.

The lack of agreement with the results of previous research work calls for some comment. Two considerations seem relevant: (a) the research of Buck and Duffy (1980) and Borod et al. (1985) was all based on numerically very small samples of brain-damaged subjects (8 and 8 patients, respectively, in the first case, and 10 and 10 in the second); (b) a second point is related to the scoring system used in preceding research (Borod et al. 1985; Kolb and Milner 1981). Scoring was done by the experimenter himself and was based on an intuitive assessment; awareness of the research hypothesis may certainly have affected scoring reliability.

A second group of results points to the lack of any possible location capable of producing selective disorders of the control of spontaneous emotional responses. This evidence may well be contradicted by future research. Nevertheless, it should be borne in mind that there are no reported clinical neurological descriptions of selective disorders in the production of spontaneous emotional responses due to cortico-subcortical lesions of the

type investigated here. On the other hand, lesions to the basal ganglia (not present in our sample) often cause disorders of spontaneous emotional communication. Returning to the question raised in the introduction, the present results are not consistent with the hypothesis of a special module which controls spontaneous facial expressiveness.
For the purpose of extending the proposed interpretation of facial control movements it could be suggested that the neurological substrate regulating a wide range of facial expressions is bilaterally and diffusely represented in the cerebral cortex.

The third set of data refers to avoidance responses. Lesions to the right hemisphere do not cause any increase in the frequency of disturbance of the production of emotional facial responses, nor any change in their quality. Nevertheless, although understanding is retained, the degree of emotional impact using objectively powerful stimuli is not as high as that observed in control subjects or left brain-damaged patients. This observation appears to be of extreme importance with regard to our understanding of the general problem of the role of the two hemispheres in managing emotional experience.
These last observations mean that it is probably necessary to make a distinction between two important aspects of emotional experience: expressive capacity, which does not seem to be lateralized, and the capacity to be in tune with the emotional content of certain stimuli, which could perhaps be related to special functions of the right hemisphere.

Final Remarks

The question underlying various hypotheses discussed in the introduction is whether the different ways in which emotional experience occurs can be thought of as depending on a single complex central "processor" which regulates all the observable manifestations, or whether there exist modes of emotional experience processing which can be broken down into smaller units.

The possibility, in brain pathology, of observing deficits which can be selectively restricted to one particular emotional sphere is consistent with the hypothetical existence of separate systems for processing experience. The absence of dissociability clearly points to the opposite interpretation. Furthermore, neuropsychological research has suggested that it is possible to identify a nonequivalent contribution by the two brain hemispheres to managing the various aspects of emotional experience.
In the present research a further study has been made of the modules assumed to be responsible for processing facial motor responses, which, ever since Darwin, have been considered universal, probably biologically determined, modes of responding to emotion-causing events and of communicating one's emotional experience to others.
One of the first important aspects to emerge has been the impossibility of locating in either hemisphere, particularly in the extensive neocortical territories, any subsystem specifically dedicated to controlling the imitative movements of the face.

The novelty of this fact must be stressed, as it clashes with the hitherto unverified assumption of a hemispheric lateralization of such a control system, following the analogy with other mechanisms for programming and regulating the body's motor responses.

Another aspect of the results is related to the processing of complex emotional facial responses, spontaneously produced under the effect of external stimuli. Also in this case, despite the large size of the brain-damaged population observed and the highly sophisticated data analysis methods used, no evidence has emerged to suggest the existence of modules located in either hemisphere.
These two aspects, as well as representing an interesting addition to our knowledge of the underlying nervous mechanisms, take on an important theoretical significance if related to the results on avoidance responses described above, which is further enhanced by those concerning the autonomic responses observed in the same patients (Caltagirone and Zoccolotti, this volume). A probably diffuse neural representation of the control of emotional and nonemotional facial responses is accompanied by a specific involvement of the right hemisphere in the control of involuntary autonomic responses and in the capacity to attach an appropriate emotional significance to cognitively understood information.

The dissociation observed during the study of brain pathology suggests the idea of a possible partition between at least two large areas of emotional experience: on the one hand, the regulation of cognitively controllable responses represented by the emotional expressions of the face and, on the other, the capacity to organize adaptive, behavioral or autonomic responses to new situations conveying emotionally relevant messages.
The contribution that neuropsychological analysis can be expected to give to our understanding of the various spheres of emotional experience and of the relationships between them could be further enhanced by an investigation of other cognitive processing skills, such as the understanding of messages having an emotional value. This could be supplemented by a study of the relationships between emotional responses to real-world stimuli and the modification of mood, which represents a more stable variation, not necessarily dependent on situations outside the subject, of emotional experience in the broader sense of the term.

References

Alford RD (1983) Sex differences in lateral facial facility; the effects of habitual functional concealment. Neuropsychologia 21: 567-570

Borod J, Caron HS (1980) Facedness and emotion in relation to lateral dominance, sex, and expression type. Neuropsychologia 18: 237-242

Borod J, Caron HS, Koff E (1981) Asymmetry in positive and negative facial expressions: sex differences. Neuropsychologia 19: 819-824

Borod J, Koff E, White B (1983) Facial asymmetry in posed and spontaneous expressions of emotion. Brain and Cognition 2: 165-175
Borod JC, Koff E, Perlman M, Nicholas M (1985) Channels of emotional expression in patients with unilateral brain damage. Archives of Neurology 42: 345-348
Bruyer R (1981) Asymmetry of facial expression in brain damaged subjects. Neuropsychologia 19: 615-624
Buck R, Duffy R (1980) Nonverbal communication of affect in brain damaged patients. Cortex 16: 351-362
Campbell R (1978) Asymmetries in interpreting and expressing a posed facial expression. Cortex 16: 473-481
Chaurasia BD, Geswami HK (1975) Functional asymmetry in the face. Acta Anatomica 91: 154-160
De Renzi E, Pieczuro A, Vignolo LA (1966) Oral apraxia and aphasia. Cortex 2: 50-73
De Renzi E, Pieczuro A, Vignolo LA (1968) Ideational apraxia: a quantitative study. Neuropsychologia 6: 41-52
Dopson WG, Beckwith BE, Tucker DM, Bullard-Bates PC (1984) Asymmetry of facial expression in spontaneous emotion. Cortex 20: 243-251
Ekman P, Friesen WV (1978) Facial Action Coding System. Palo Alto, CA: Consulting Psychologists Press
Ekman P, Hager JC, Friesen WV (1981) The symmetry of emotional and deliberate facial actions. Psychophysiology 18: 101-106
Etkoff NL (1986) The neuropsychology of emotional expression. In Goldstein G, Tarter E (eds) Advances in clinical neuropsychology. Plenum, New York
Gainotti G, Miceli G, Caltagirone C (1977) Constructive apraxia in left brain damaged patients: a planning disorder? Cortex 13: 109-118
Geschwind N (1975) The apraxias: neural mechanisms of disorders of learned movements. Am Sci 63: 188-195
Hager JC, Ekman P (1985) The asymmetry of facial actions is inconsistent with models of hemispheric specialization. Psychophysiology 23: 307-318
Heilman KM, Schwartz HD, Watson RT (1978) Hypoarousal in patients with neglect and emotional indifference. Neurology 28: 229-232
Koff E, Borod J, White B (1981) Asymmetries for hemiface size and mobility. Neuropsychologia 19: 825-830
Kolb B, Milner B (1981) Performance of complex arm and facial movements after focal brain lesions. Neuropsychologia 19: 491-503
Lynn JG, Lynn DR (1938) Face-hand laterality in relation to personality. J Abnormal Soc Psychol 33: 291-322
Morrow L, Vrtunski B, Kim Y, Boller F (1981) Arousal responses to emotional stimuli and laterality of lesion. Neuropsychologia 19: 65-71
Moscovitch M, Olds J (1982) Asymmetries in spontaneous facial expressions and their possible relation to hemispheric specialization. Neuropsychologia 20: 71-81
Natham PW (1947) Facial apraxia and apraxic dysartria. Brain 70: 449-478
Rinn WE (1984) The neuropsychology of facial expression: a review of the neurological and psychological mechanisms for producing facial expressions. Psychol Bull 95: 52-77

Ross ED, Mesulam MM (1979) Dominant language functions of the right hemisphere? Prosody and emotional gesturing. Arch Neurol 36: 144-148
Rubin DA, Rubin RT (1980) Differences in facial asymmetry between left and right-handed children. Neuropsychologia 16: 437-481
Sackeim HA, Greenberg MS, Weiman AL, Gur RC, Hungerbuhler JP, Geschwind N (1982) Hemispheric asymmetry in the expression of positive and negative emotions: neurologic evidence. Arch Neurol 39: 210-218
Sackeim HA, Gur R, Saucy MC (1978) Emotions are expressed more intensely on the left side of the face. Science 202: 434-436
Schwartz GE, Ahern GL, Brown SL (1979) Lateralized facial muscle response to positive and negative emotional stimuli. Psychophysiology 16: 561-571
Strauss E, Wada J, Losaka B (1983) Spontaneous facial expressions occurring at the onset of focal seizure activity. Arch Neurol 40: 545-547
Zoccolotti P, Caltagirone C, Benedetti N, Gainotti G (1986) Perturbation des responses vegetatives aux stimuli emotionnels au cours des lesions hemispheriques unilaterales. L'Encephale 12: 263-268
Zoccolotti P, Scabini D, Violani C (1982) Electrodermal responses in patients with unilateral brain damage. J Clin Neuropsychol 4: 143-150

Autonomic Reactivity and Facial Expression of Emotion in Brain-Damaged Patients

Carlo Caltagirone[1], Pierluigi Zoccolotti[2], Giancarlo Originale[2], Antonio Daniele[3] and Alessandra Mammucari[2]

[1] Neurologic Clinic, II University of Rome "Tor Vergata", Via O. Raimondo, 00173 Rome, Italy
[2] Department of Psychology, University of Rome "La Sapienza", Via degli Apuli 8, 00185 Rome, Italy
[3] Neurologic Clinic, Catholic University of Rome, Largo A. Gemelli 1, 00100 Roma

INTRODUCTION

The relationship between emotional behaviour and unilateral brain damage has aroused growing interest since the first systematic observations made by Gainotti (1969, 1972). In a specially planned investigation carried out on a large population of brain-damaged patients it was found that those affected by lesions of the left hemisphere often also had fits of anxiety, crying and despair (defined as "catastrophic" reactions) while those with lesions of the right hemisphere frequently displayed reactions involving fatuous playfulness, a tendency to minimize their nevertheless serious symptoms and behaviour characterized by indifference. This confirmed previous clinical evidence (Critchley 1955; Hecaen 1962) and represented a significant step forward when it was suggested that a relationship existed between the distinct emotional expressions of the various groups of brain-damaged patients suffering from specific cognitive disorders resulting from lesions to either hemisphere: unilateral spatial agnosia and anosognosia were found to be related to the display of indifference in those with right brain damage, while the seriousness of the aphasic disorder was related to the existence of catastrophic reactions in those with left brain damage. Gainotti (1972) suggested interpreting the differences found in emotional behaviour in terms of the relative appropriateness of the response. The catastrophic responses by the left brain-damaged patients were interpreted as a dramatic but nevertheless appropriate response to the acknowledgment of one's illness, while the display of indifference by right brain-damaged patients was attributed, in view of its inappropriateness, to a nuclear disorder in the organization of emotions itself.

The fact that the two brain hemispheres could make a differential contribution to the organization of emotional behaviour had already emerged from studies based on the unilateral injection into the carotid of sodium amytal (Alemà and Donini 1960; Terzian and Cecotto 1959). The temporary inactivation of one hemisphere obtained under the effect of this drug produced results similar to those observed by Gainotti (1969, 1972) in brain-damaged subjects.

Experimental Brain Research Series 18

Therefore, starting in the early '1970s, there was a steady increase in research based on the hypothesis of a differential specialization of the two brain hemispheres in regulating emotional behaviour. Nevertheless, it must be pointed out that the numerous researchers working in this field were mainly concerned with studying the components related to understanding emotion-producing stimuli, focusing their attention on the capacity of the different groups of brain-damaged patients to recognize the emotional value of stimuli presented in different perceptual forms (Boller et al. 1979; Brownell et al 1983; Cicone et al. 1980; DeKosky et al. 1980; Gardner et al. 1975; Heilman et al. 1975). A smaller number of investigations were made of spontaneous emotional behaviour in the different groups of brain-damaged patients. Sackeim et al (1982), in a review of cases of emotional outburst ascribable to single hemisphere brain damage or to critical phenomena of an epileptic nature, found a relationship between type of emotions and brain hemisphere. The greater frequency of laughter episodes among epileptics with left hemisphere focal disorders ran parallel to the greater frequency of crying fits in patients with right hemisphere epileptic disturbances. When the consequences of lesions to one hemisphere were examined the picture was a reversed image of the above: right brain-damaged patients displayed more frequent fits of spastic laughter, while spastic crying fits were more common among left brain-damaged patients. On the basis of the opposite effects caused by lesions and by hyperactivation phenomena, Sackeim et al (1982) hypothesized that the two brain hemispheres shared the capacity to express and understand emotions according to their valence: "positive" emotions were recognized and expressed more readily by the left hemisphere, while "negative" emotions were processed better by the right hemisphere.

Starting from the observation of patients affected by dysprosody (i.e., with disturbance of stress, pitch and rhythm of speech), Ross and coworkers (Ross 1981, 1984; Ross and Mesulam 1979) claimed that the right hemisphere plays a primary role in the organization of the extra-linguistic (i.e., affective-prosodic) aspects of communication and that the anatomico-functional organization of this system was a mirror image of the organization of propositional language in the left hemisphere. Cook (1984) suggested a neurophysiological basis for this specialization of functions, pointing out that the activation of the language centers of the left hemisphere causes the inhibition of the contralateral areas (homotopic callosal inhibition) and an activation of the neighbouring areas. In this hypothesis the analytical-linguistic processing that takes place in the left hemisphere is causally linked to the contextual metalinguistic analysis performed by the right hemisphere. Disturbance of the system of extra-verbal communication, which frequently occurs during right hemisphere damage, would account for the difficulties encountered by these patients in expressing emotions. According to Ross's interpretation the characteristic display of indifference is only apparent, and is the result of the difficulty experienced in communicating emotional response. The actual emotional experience is unaffected.

These interpretative hypotheses have aroused the interest of workers in the behavioural field and have more or less explicitly inspired a large body of research based on experimental models using information available from the study of normal subjects. Most of the work, however, has been addressed to examining the processes involved in understanding stimuli of emotional significance. Only a small amount of research has been aimed at evaluating possible variations in emotional experience as a result of brain damage. These studies include those involving autonomic response to emotion-producing stimuli and those dealing with the production of emotional facial expressions.

Autonomic Reactions in Patients with Unilateral Brain Damage

One of the first studies of autonomic response in brain-damaged patients was made by Heilman et al. (1978), who used skin conductance level and response as indicators of autonomic activation. During painful stimulation produced by slight electric shocks, patients with right parietal brain damage and hemi-neglect, displayed a lower reactivity on both indicators compared with controls. On the other hand, patients with left parietal lesions and aphasia exhibited increased skin conductance responses compared with controls.

Morrow et al. (1981) showed that skin conductance responses by control subjects to images with strong emotional valence (e.g. a hand with a conspicuous bleeding cut) were greater than for analogous responses to stimuli without any emotional content. The same response pattern was present, although somewhat reduced, in patients with lesions to the left hemisphere. In right brain-damaged patients no difference in autonomic response was observed between the two stimulus situations. Moreover, unlike the investigation of Heilman et al. (1978), in the latter case the differences between the various groups considered were found only in the case of skin conductance responses, while no difference was observed in conductance levels.

These results were confirmed in subsequent research (Zoccolotti et al. 1982). The patients with left hemisphere lesions exhibited skin conductance responses to emotion-arousing slides for the whole duration of the test. On the other hand, right brain-damaged patients displayed a clearcut electrodermal response only to the first (neutral) stimulus, followed by a rapid adaptation which was independent of the emotional valence of the stimulus. The existence of a response to the initial stimulus seems to exclude the likelihood that the electrodermal system is simply inactive in patients with right hemisphere lesions and instead points to the maintenance of the orientation response to new stimuli. In the same investigation, Zoccolotti et al. (1982) gave their subjects a task involving the recognition of emotional facial expressions. In keeping with the idea of a reduced capacity for processing emotion-arousing stimuli (e.g., Cicone et al. 1980), right brain-damaged patients performed worse than left brain-damaged patients. Nevertheless, the differential pattern of autonomic responses remained the same even after covariation for the performance obtained in this task. These results suggest that the absence of a differentiated autonomic response in right brain-damaged patients is more or less independent of their perceptual analysis deficit (Zoccolotti et al. 1982). Results indicative of dissociation in right brain-damaged patients between disturbance of understanding the significance of visual stimuli and the clinically observed pattern of indifference have also been reported by Etcoff (this volume).

Overall, the above investigations point to a disturbance of autonomic responses to emotion-arousing stimuli in right brain-damaged patients. However, there is a certain variability in the results of this group of investigations. Neither the work of Morrow et al. (1981), nor that of Zoccolotti et al. (1982), has clearly confirmed the difference found by Heilman et al. (1978) in skin conductance levels and interpreted by them in terms of a reduction, in right brain-damaged patients, of the level of basal tonic activation. Furthermore, it is not clear what role is played by left brain damage: in the work of Heilman et al. (1978), patients with a lesion of this hemisphere exhibited increased skin conductance compared with controls, while in the work of Morrow et al. (1981) this pattern is reversed. However, it seems worth stressing that the generalization of the conclusions is, in any case, limited by the use of a single autonomic activation parameter (skin conductance).

Facial Expressiveness in Patients with Unilateral Brain Damage

A first systematic investigation (Buck and Duffy 1980) indicated a lower degree of expressiveness in patients with right brain damage than in patients with left brain damage and controls. Borod et al. (1985) also obtained similar results while analysing spontaneous facial actions in brain-damaged patients. They pointed out more specifically that patients with right frontal lesions display the greatest disturbances. Contradictory results were obtained by Kolb and Milner (1981). Their investigation, based on observation of spontaneous facial actions during standard neuropsychological procedures, indicated that it was the presence of a frontal lesion on either side which produced the greatest disturbance in facial behaviour. It must be stressed that all the above investigations were based on subjective observer evaluations of the appropriateness and the intensity of facial expressions. It is also not clear whether observers were aware of the underlying hypotheses (e.g. Borod et al. 1985). Finally, one must consider the complexity of any such evaluation in the comparatively frequent case of facial hemiparesis in which facial motility is altered both statically and dynamically.

Recently Ekman and Friesen (1978) developed a system (Facial Action Coding System, FACS) for evaluating facial expressions on the basis of an analytical examination of each possible unit of facial action. When used by an expert scorer, this system affords a comparatively objective evaluation, that is one that is independent of observer bias, and of the nature and intensity of the facial expression observed. Furthermore, this analytic method seems to be particularly well suited to the case of patients with brain damage involving the facial cranial nerves, as is frequently the case, since it allows an accurate overall evaluation to be made of the emotion expressed on the basis of only one hemisphere.

Within the framework of the research reported herein, this technique was used to ascertain possible variations in the type and intensity of facial expressions in unilateral brain-damaged patients subjected to emotion-arousing stimuli. The results show that there is no difference in function of the damaged hemisphere with regard to any of the parameters considered (Pizzamiglio and Mammucari, this volume).
It may thus be concluded that evidence in support of a specific role of the right hemisphere in mediating the production of emotional facial expressions is contradictory. Whether or not it is possible to infer from studies of spontaneous expressiveness in brain-damaged patients that there exists a difference between hemispheres, it still remains to be clarified whether the greater difficulty experienced by right brain-damaged patients in such activities is due to a dissociation between the capacity for emotional communication and the interior experience of such emotions, as is explicitly postulated by Ross (1984), or whether the weaker expressive activity is rather the consequence of a disorganization of the emotional experience itself, as would seem to be suggested by the results of research on autonomic responses in brain-damaged patients (Heilman et al 1978; Morrow et al. 1981; Zoccolotti et al. 1982).

EXPERIMENTAL RESEARCH

The experimental research described herein was based on the above considerations. Our research is part of a research program, the overall aim of which is to examine multiple, integrative parameters of emotional behaviour in a large sample of brain-damaged patients in such a way as to systematically monitor aspects of facial expression (both spontaneous and posed), the related autonomic mechanisms and the understanding of emotions. The theoretical framework and the overall plan of the research, together with some of the results obtained are described in other chapters (Pizzamiglio and Mammucari, this volume; Gainotti, this volume). The specific aspects of the work described in detail below refer to the study of different autonomic parameters and their possible relations with spontaneous emotional facial expressions.

As has been seen above, investigations of autonomic responses in brain-damaged patients has been limited to the analysis of electrodermal activity alone.

The use of more than one autonomic indicator is important in that autonomic patterns indicating a "directional fractionation" (e.g., Lacey 1967) have frequently been observed during the administration of suitable stimuli. In the case of a subject exposed to external stimuli, there is a simultaneous activation of sympathetic parameters such as skin conductance responses or pupil dilation, accompanied by changes in autonomic parameters indicative of an activation of the parasympathetic system such as falling arterial pressure and pulse rate. It has been observed that heart rate deceleration is enhanced in the case of stimuli with negative emotional valence (Libby et al. 1973). In the same investigation, stimuli with positive emotional valence were found to produce responses midway between negative and neutral ones.

The combined examination of autonomic parameters and facial expressions seems to be particularly important in the light of the hypothesis of facial feedback originally proposed by Darwin (1872). An analytical description of evidence for and against this hypothesis has been made by Lanzetta and McHugo (this volume). Briefly, the hypothesis states that the feedback from a facial expression causally modifies the way an emotion is experienced. It follows that the actual presence/absence of facial expression or even its intensity can modify the emotional experience at any given moment. Various studies in which autonomic parameters are taken as indicators of a subject's emotional involvement have indicated a positive correlation between the intensity of facial responses and that of the associated autonomic responses (Lanzetta et al. 1982; Lanzetta et al. 1976; Kleck et al. 1976): subjects explicitly instructed to control their facial expressions or induced to do so by peculiar environmental conditions (e.g. the presence of an observer) display weaker skin conductance responses to a nociceptive stimulus than the control subjects. The mutual interdependence of autonomic and facial factors has also been stressed in research by Ekman et al. (1983). In a sample consisting of trained subjects (professional actors and facial behaviour researchers), specific autonomic responses patterns were produced for different emotional expressions, independently of the way they were evoked (i.e., posed or spontaneous). Previous analyses (Pizzamiglio and Mammucari, this volume) have shown that spontaneous facial expression is not differentially impaired by lesions in either hemisphere. However, the possibility can be raised that a brain lesion may have the effect of dissociating facial expressivity from emotional experience, as indexed by autonomic arousal. Examination of the video recordings revealed that numerous subjects made various head and eye movements which enabled them to avoid looking at the screen for many seconds at a time (Pizzamiglio and Mammucari, this volume). In agreement with the hypothesis of a response of an emotional nature, gaze aversion occurred most frequently during the projection of the negative film segment. Interestingly, this parameter significantly distinguished between the patient's groups: in the case of the negatively charged film, unlike both LBD and controls, very few RBD patients showed any evidence of gaze aversion away from the screen.

The presence of active avoidance movements is interesting from a psychophysiological standpoint since it is well known that active coping in the case of stressful situations is an important factor modulating cardiovascular responding (Obrist, 1976). In situations of passivity heart rate deceleration is observed; if the individual is given the opportunity of taking action that may permit avoidance of the stimulus, heart rate acceleration is commonly observed (Elliott 1969; Light and Obrist 1980; Malcuit 1973; Sandman 1975). However, the effect is not directly related to the amount of muscular activity. In fact, greater heart acceleration was observed in a situation of active avoidance than in one of passivity, even in the presence of identical movements (Light and Obrist 1980). In general, the heart deceleration is seen as a part of the freezing reaction shown by animals when no way of escape is given (Sandman 1975). Consistently, it is shown only by those species, such as cats, rabbits and man, that display such a reaction. Conversely, other species, such as monkey or pigeon, who do not show a freezing reaction, generally display acceleratory cardiac responses to unavoidable noxious stimuli (Obrist 1976). The effect of the possibility of avoiding the stressful stimulus is specific of cardiovascular functioning. No differential effect of the possibility of avoiding is observed in the case of a unidirectional parameter such as skin conductance (Bandler et al. 1968; Sandman 1975). These considerations suggested that it might be interesting to look at whether the observed patterns of gaze aversion to stressful stimuli are associated or not with variations in cardiovascular responding. It was felt that this might help shed light on the nature of the autonomic flattening displayed by RBD patients.

In conclusion, therefore, the combined study of several autonomic indicators and of facial expressiveness in brain-damaged subjects appeared to afford a particularly sensitive paradigm for measuring possible affective alterations in the light of the hypotheses previously advanced on the role of the two brain hemispheres in the organization of emotional behaviour.

The sample studied comprised 40 patients with unilateral brain damage, half of the right hemisphere and half of the left, plus 20 control subjects. The diagnosis of damage to the hemisphere was based on clinical evidence, EEGs and neuroradiological examination (in most cases a brain CAT scan had been made). Left-handed and ambidextrous subjects were excluded, together with patients whose lesions were too acute or serious to allow them to be accompanied from the Neurology Department to the Psychophysiology Laboratory where all the experimental work was done. One final exclusion criterion used was the existence of disturbances of heart rate. There were no significant inter-group differences with regard to age, education and relative number of males and females. In 35 patients the lesion had a vascular etiology and in five (three right and two left brain-damaged patients) a tumoral etiology.

The patients were led to a comfortable, acoustically insulated room, where they were asked to sit in an armchair in conditions of silence and dim lighting. Film segments of varying emotional intensity were projected on a screen while the face of the subject was filmed using a closed-circuit television system. In order to familiarize the subject with the experimental situation, four film segments varying in content were shown which were not subsequently processed. Then, without any break in the continuity, the stimuli used to arouse emotional reactions were projected. The latter consisted of short film segments presented in the following order: the surgical cleaning of a serious burn (sequence with negative connotations), sea waves breaking monotonously on the rocks (sequence with no emotional connotations) and a puppy playing with some flowers giving rise to a series of amusing situations (sequence with positive connotations).

The electrodes used to record the psychogalvanic reflex were applied to the fingers of the hand homolateral to the brain lesion in the subjects belonging to the experimental group, and to the right hand in the case of the control subjects. Basic electrodermal activity was taken to be the basal value of skin conductance (in $\mu mho/cm^2$) recorded just prior to the projection of the three film segments. The skin conductance response was taken to be the maximum increase in conductance occurring during the first 20 s of projection of each sequence. The responses were converted into units of quadratic modification of conductance using the formula $\sqrt{\Delta}$ conductance x 1000 suggested by Lubin and Johnson (1972) and used in previous research (Morrow et al. 1981; Zoccolotti et al. 1982). The variations in heart rate were measured using a cardiotachometer and taking as the base value the mean number of interbeat intervals recorded in the 10 s preceding the projection of each film segment, and comparing this value with that recorded during the first 20 s of projection of the same sequence.

At the end of each experimental session the subjects were interviewed to ascertain their understanding of the film sequence they had seen. Only those subjects who proved to have a good overall understanding of the film segments were included in the experiment.

In accordance with the procedures described by Ekman and Friesen (1978), the recordings of individual patients were examined frame by frame in order to identify the individual facial actions and situate them precisely in the context of film viewing. Using this evaluation method it was possible to identify facial action patterns which can be considered as more or less coherent and unambiguous manifestations of one of the six fundamental emotions (happiness, sadness, disgust, surprise, fear and anger).

The data concerning spontaneous emotional facial expression aroused by the film segments projected are presented analytically by Pizzamiglio and Mammucari (this volume). They show that no differences occur for any of the films in the number of right or left brain-damaged patients displaying facial expressions appropriate to the affective connotation of the film. In contrast, the presence of gaze aversion in the negative film segment was related to the side of the lesion: controls and left brain-damaged patients showed significantly more avoidance behaviours than right brain-damaged patients.

In examining the possible association between facial expressions and avoidance movements, and autonomic reactivity, it was deemed important to ascertain whether they were independent of each other, or whether the presence of one facilitated or inhibited the presence of the other. For this purpose, on the basis of the number of times in which facial emotional expression behaviour (.22) and avoidance behaviour (.32) occurred, the probability of an association occurring between the two behaviours in the case in which the two expressive events were independent of each other was calculated (.07). It was thus possible to determine that the frequency with which this association was actually observed (.07) did not differ from the theoretical probability of its occurring by chance ($X^2 < 1$, n.s.). The basic independence of the two expressive behaviours was also confirmed by an

examination of the video recordings referring to the order in which the two behaviours appeared in time. This analysis showed that whenever either appropriate emotional responses or avoidance behaviour were displayed by the patients they did not follow any fixed sequence.
Analysis of data referring to the sample as a whole and neglecting facial expressions showed that the negative film produced significant variations in the extent of both the modifications of heart rate and of skin conductance compared with the neutral one. On the other hand, no difference could be detected in either of these parameters between the positive and the neutral film sequences. It should be stressed that this result is specific to autonomic responses since a clearcut distinction was made between the positive and the neutral film in terms of appropriate facial expression. It is significant in this connection that the autonomic responses are usually more pronounced when the emotional stimuli are negative than when they are positive (Lanzetta and McHugo, this volume). However, the use of only two films prevents us from ruling out the possibility that it is actually wholly or partly a difference in intensity between the two stimuli used. In any case, for this reason, further processing was applied only to the data referring to the negative and neutral films, respectively. In particular, the results described are based on four separate analyses of variance. In one case, the heart rate (expressed as mean intersystolic interval) was analysed as a function of four factors: patient group (left brain-damage/right brain-damage/controls) and emotional facial expressions as a response to film type (presence/absence) as non repeated factors, with treatment (baseline/film) and film type (neutral/negative) as repeated factors. A second analysis was the same except that the second factor consisted of the presence/absence of avoidance movements. Two analyses, one concerning the presence of facial expressions and the other that of avoidance movements were also carried out using skin conductance response as a dependent measure. These analyses were the same as the preceding ones except for the absence of the treatment factor (the maximum phasic response during the film was considered neglecting the pre-stimulus baseline).

Modification of Heart Rate

The variations in heart rate in the three groups of subjects occurring during the screening of the negative and the neutral film segments are shown in Fig 1. As expected, comparison between the baseline heart rate and that recorded during film projection revealed modifications which were indicative of an increase in the mean intersystolic interval (i.e., a slower heart rate).

Furthermore, the deceleration response varied according to the film sequence. It was greater for the negative film than for the neutral one. The most important result for our hypothesis is the fact that the controls and left brain-damaged patients produced greater deceleration responses when exposed to negative emotionally-charged film stimuli, while the right brain-damaged patients generally produced smaller deceleration responses without any particular differences due to film type. This result is supported by the

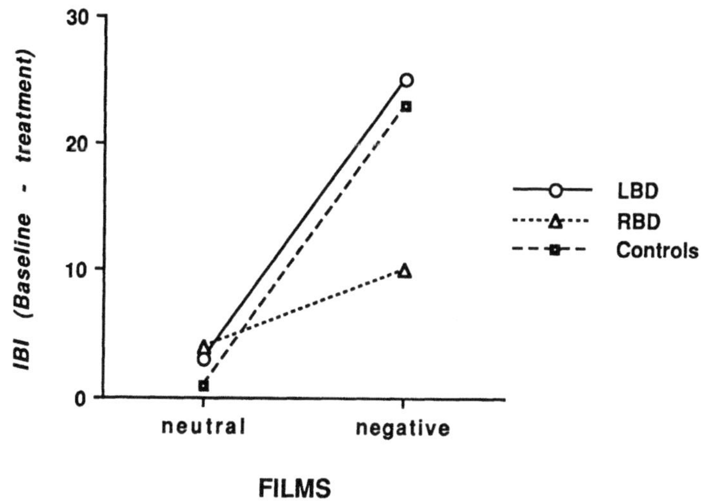

Fig. 1 Differences between mean interbeat intervals (IBI) recorded during the baseline period and during the films. Highest values indicate highest cardiac decelerations. Data obtained from controls, left brain-damaged (LBD) and right brain damaged (RBD) patients are separately shown as a function of the neutral and negative film

existence of an interaction between patient group, treatment and film type ($F = 3.15$, $p < 0.05$).

Modification of Electrodermal Activity

Figure 2 shows how skin conductance responses in control subjects and patients with left brain-damage are, on average, higher than those exhibited by patients with right brain-damage (main effect of patient group factor; $F = 4.4$, $p < 0.05$).

Furthermore, the values obtained in response to negative emotionally charged film segments are greater than those for neutral films (main effect of the film type factor; $F = 9.04$, $p < 0.001$). The most interesting aspect is that this effect varies for each of the three groups of patients: the mean responses of right brain-damaged patients show no differences linked to film type; on the contrary, control subjects and left brain-damaged patients display greater conductance responses for the emotionally charged film than for the neutral one (patient group interaction with film type; $F = 2.5$, $p < 0.05$).
No difference among the three groups was found for skin conductance values preceding stimulus presentation. This result confirms those previously obtained using similar populations (Morrow et al. 1981; Zoccolotti et al. 1982).

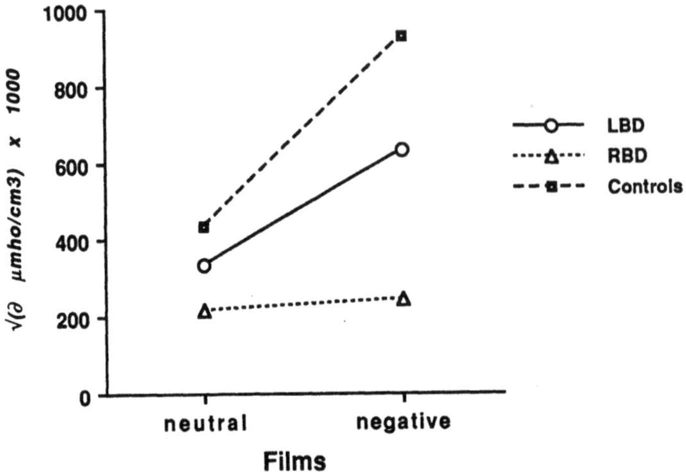

Fig. 2 Skin conductance responses recorded during neutral and negative film. Data obtained from controls, left brain-damaged (LBD) and right brain damaged (RBD) patients are separately shown

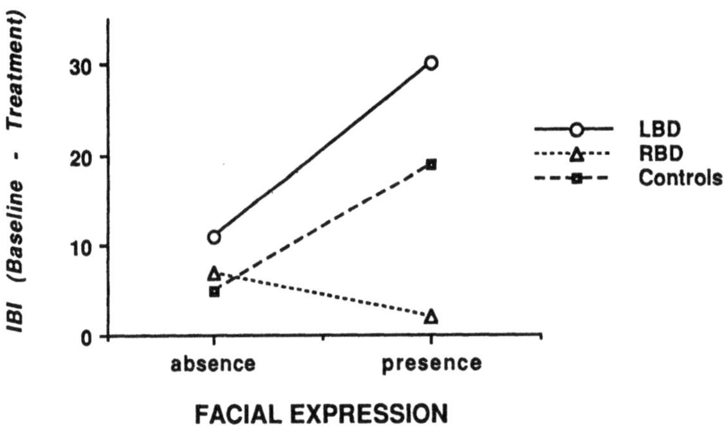

Fig. 3 Differences between mean interbeat intervals (IBI) recorded during the baseline period and during the films. Highest values indicate highest cardiac decelerations. Data obtained from controls, left brain-damaged (LBD) and right brain damaged (RBD) patients are separately shown as a function of the presence of facial emotional expressions to the negative film

Modification of Heart Rate as a Function of Presence of Spontaneous Emotional Facial Expressions

In Fig 3 the variation in heart rate for the three groups is shown as a function of the presence of facial expressions.

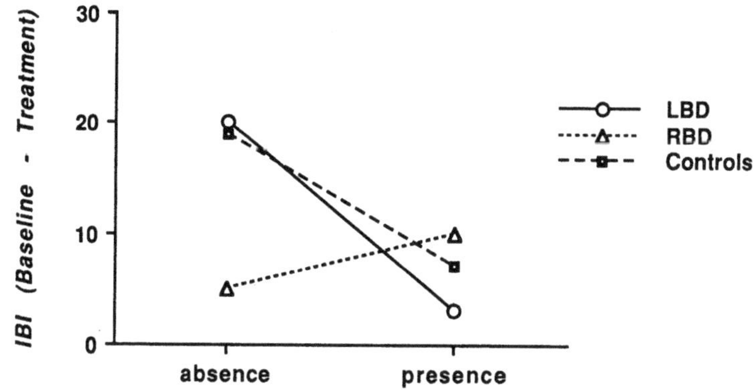

Fig. 4 Differences between mean interbeat intervals (IBI) recorded during the baseline period and during the films. Highest values indicate highest cardiac decelerations. Data obtained from controls, left brain-damaged (LBD) and right brain damaged (RBD) patients are separately shown as a function of the presence of avoidance behaviour during the negative film

The heart rate deceleration response produced by controls and left brain-damaged patients on exposure to the film stimulus are greater in those patients who have produced emotional facial expressions appropriate to the connotation of the film. Right brain-damaged patients, on the other hand, exhibit no modification of the heart rate parameter whether or not they produce an emotional facial expression appropriate to the film. This set of results is supported by the presence of an interaction between patient group, presence/absence of facial expressions, and treatment ($F = 2.9$, $p = 0.06$).

Modification of Heart Rate as a Function of Presence of Avoidance Behaviours

Variations in heart rate as a function of presence/absence of avoidance behaviour during the projection of the negative film are shown in Fig 4, which also illustrates the interaction between groups, treatment and presence/absence of avoidance behaviours ($F = 2.9$, $p = 0.06$).

The comparison of baseline and film values shows greater heart rate deceleration for subjects who watched the entire film sequence than for subjects who displayed avoidance behaviour. This effect was found for both control subjects and left brain-damaged patients. In right brain-damaged patients the presence or absence of avoidance behaviour was not related to differential cardiac responses.

Variation in Skin Conductance as a Function of the Presence of Facial Expressions and Avoidance Behaviour

In the case of skin conductance responses analysis of variance indicated no main effect or any interaction linked to the factor of the presence/absence of emotional facial expressions. The same result emerged during the analysis to study the effect of avoidance behaviour.

DISCUSSION

Overall the results indicate a disturbance of autonomic reactivity in patients with right hemisphere brain damage. It should be noted that the differences found specifically refer to visceral variations depending on the affective connotation of the stimuli. Consequently, no difference was observed in the baseline recordings preceding exposure to the stimuli in the three groups considered, either in skin conductance or in heart rate. Furthermore, both the control subjects and the left and right brain-damaged patients reacted to the complex of stimuli presented according to a pattern of psychophysiological responses in agreement with what Lacey (1967) termed as "directional fractionation": i.e., they exhibited both an increase in skin conductance (ascribable to activation of the sympathetic system) and a reduction in heart rate (ascribable to activation of the parasympathetic system). These results seem to indicate that right brain-damaged patients are not characterized by a disturbance of the autonomous nervous system (ANS), or by any incapacity to implement attentional mechanisms on exposure to new external stimuli. These observations are consistent with the evidence that the orientation response is preserved in these patients (Zoccolotti et al. 1982).

A clearcut distinction can, however, be observed between right and left brain-damaged patients with regard to variations in the autonomic parameters as a function of the affective connotations of the film segments: both left brain-damaged patients and controls displayed maximum deceleration of heart rate for negative visual stimuli and minimum, although some, response to stimuli with no apparent affective connotations. Furthermore, a parallel differentiation was observed for skin conductance responses. These results are consistent with the observations made for similar stimuli by Libby et al. (1973) for normal subjects. On the other hand, right brain-damaged patients displayed no difference between emotional stimuli and neutral ones in the case of skin conductance and heart rate. It must be stressed that the absence of variation of autonomic responses with emotional stimuli cannot be explained in terms of the incapacity of the right brain-damaged patients to appreciate the meaning of the stimuli insofar as all subjects were able to accurately describe the content of the film segments viewed.

It can be concluded that right brain-damaged subjects exhibit selective disturbance of the autonomic responses to emotional stimuli. In particular, the evidence confirms the findings of previous investigations (Heilman et al. 1978; Morrow et al. 1981; Zoccolotti et al. 1982) and points to a disturbance in electrodermal reactivity in patients with lesions to the right hemisphere. Moreover, the existence of a similar difference in the heart rate indicates that the disturbance affecting these patients is neither specific to a single autonomic indicator nor to a single subdivision (sympathetic or parasympathetic) of the ANS. The general nature of this conclusion suggests the hypothesis that the absence of differential autonomic arousal plays a comparatively important role in the onset of alterations to emotional behaviour revealed by the clinical examination of patients with right hemisphere lesions.

Another interesting feature of our research is the study of the autonomic responses as a function of different expressive behaviours (emotional facial expressions, avoidance behaviour). In agreement with the facial feedback hypothesis (Lanzetta and McHugo, this volume) the presence of emotional facial expressions was found to be positively related to the size of the cardiac response in the control and left brain-damaged subjects. In other words, in these two groups, those that produced facial responses were the same as those displaying the greatest variations in heart rate. This pattern was not confirmed for skin conductance responses.

Also the avoidance behaviour was accompanied by differential patterns of heart rate variation. In the control and left brain-damaged subjects, the absence of head and eye movements intended to avoid looking at the film sequence with negative connotations was related to a stronger cardiac response. It is interesting to observe that this result is consistent with those obtained in systematic studies of the role of avoidance behaviour in modulating visceral responses to adverse or negative stimuli (Elliott 1969; Light and Obrist 1980; Malcuit 1973; Sandman 1975). It has been suggested that avoidance acts by actively reducing the experience connected with the stressful stimulus (Malcuit 1973). Also in this case the patients with right hemisphere lesions differ from the other two groups considered. On the one hand, only a few of them display any avoidance behaviour; on the other, although the majority of the patients seem to watch the negative film sequence attentively and without interruption, they display no accompanying heart rate deceleration. It should be emphasized that the relationship between avoidance behaviour and autonomic modifications refers only to the cardiac parameter. The absence of any relationship between skin conductance and avoidance behaviour is consistent with the results of previous studies with normal subjects (Bandler et al. 1968; Sandman 1975).

A word of caution must be said about the interpretation of the complex pattern observed. The results indicate that the presence of emotional facial expressions is independent of that of avoidance behaviour. This would therefore justify the decision to analyse these two behaviours separately. However, it is also clear that the small size of the sample studied prevents any inter-subject differences related to both behaviours being taken into consideration. The results reported clearly show, however, how important it is to take an integrated view of the expressive behaviour and autonomic reactivity in evaluating emotional alterations in patients with unilateral brain lesions. In particular, it would seem useful to plan experimental situations so as to be able to systematically control both the possibility of avoiding the adverse stimulus and the possibility of facilitating or inhibiting facial expressiveness.

Overall the results of the research seem to point to a situation which is only partly in agreement with the above theories concerning the relation between emotional behaviour and unilateral brain lesions.

The hypothesis that the right hemisphere is dominant in extra-verbal communication as the left hemisphere is dominant in linguistic communication (Ross 1984) is not supported by our experimental data. According to the above hypothesis right brain-damaged patients would be expected to display substantially intact emotional experience, as indicated by the autonomic parameters, while their characteristic indifference is ascribable to the difficulty they experience in understanding and expressing emotions. Our research indicates, on the contrary, that the facial expressiveness of patients with lesions to the right hemisphere is comparable to that of left brain-damaged patients while they display a generalized disturbance of autonomic reactivity.

The hypothetical interaction between differential connotative aspects of the emotions and the hemisphere with the lesions (Sackeim et al., 1982) is neither confirmed nor explicitly contradicted by our data. Indeed, the film sequence used to elicit positive emotional responses proved to be indistinguishable from the neutral one for all the autonomic parameters examined, although it differed from it in terms of emotional facial expressions. However, it must be pointed out that the latter parameter also did not indicate any differences between right and left brain-damaged subjects in the case of the positive film sequence; this latter evidence does not favour a selective compromising of emotional tasks depending on which hemisphere is damaged.

Our evidence tends rather to point to a disturbance of the autonomic responses and their relations with expressive behaviour only in patients with right brain damage. These results partially agree with Gainotti's (1972) hypothesis. He claimed that the right hemisphere plays a role in the modulation of emotional behaviour in general. However, in our research differences were observed in patients with different brain lesions only with regard to the autonomic parameters and the frequency of avoidance behaviour. Indeed, patients with right hemisphere damage did not differ from patients with left brain-damage in the frequency of their facial expressions. These differences between the various parameters of emotional behaviour seem to suggest that the right hemisphere plays a more selective role in modulating such behaviour. In particular, the results point to a distinction between processes regulating the expression of emotions and processes controlling motor responses associated with the emotional experience. On the basis of the latter an individual could, depending on the situation and on his personality, respond to a stressful event by attacking the source of the danger or by escaping from it. The sum total of these responses identifies the individual's capacity for converting his emotional experience into action and has been defined as a "coping reaction" (Lazarus 1969). The absence of any difference in facial expressiveness between the hemispheres suggests that the right hemisphere does not play any particular role in generating a differential emotional reaction to the stimulus or in its communication, at least through the face. Conversely, it seems reasonable to hypothesize that this hemisphere plays a role in the modulation of the processes supporting the production of the coping reaction. Patients with lesions of the right hemisphere would thus not have a diminished capacity for appreciating and communicating the emotional value of a stimulus, but would be incapable of implementing motor behaviour appropriate to the emotional experience (e.g. fight or flight responses) and would not display any of the modifications of the autonomic parameters associated with them.

This hypothesis may be of use in interpreting the clinical description of these patients. As we have seen, in conditions of clinical examination they often behave with indifference or even fatuousness (Gainotti 1969, 1972). Gainotti (1972) has stressed the fact that these behaviours are characterized in particular by affective inconsistency. In terms of the above hypothesis, the indifference of right brain-damaged patients is the expression of a disturbance of the mechanisms through which the individual responds to the stressful event by means of a reaction in harmony with his emotional experience.

REFERENCES

Alemà G, Donini G (1960) Sulle modificazioni cliniche e EEG da introduzione intracarotidea di iso-amil-etil-barbiturato di sodio nell'uomo. Boll Soc Ital Biol Sper 36: 900-904

Bandler R J, Madaras G R, Bem D J (1968) Self-observation as a source of pain perception. J Pers Soc Psychol 9: 205-209

Boller F, Cole M, Vrtunski P B, Patterson M, Kim Y (1979) Paralinguistic aspects of auditory comprehension in aphasia. Brain Lang 7: 164-174

Borod JC, Koff E, Perlman M, Nicholas M (1985) Channels of emotional expression in patients with unilateral brain damage. Arch Neurol 42: 345-348

Brownell H H, Michel D, Powelson J, Gardner H (1983) Surprise but not coherence: sensitivity to verbal humour in right hemisphere patients. Brain Lang 18: 20-27

Buck, R Duffy, R (1980) Nonverbal communication of affect in brain damaged patients. Cortex16: 351-362

Cicone M, Wapner W, Gardner H (1980) Sensitivity to emotional expressions and situations in organic patients. Cortex 16: 145-158

Cook N D (1984) Homotopic callosal inhibition. Brain Lang 23: 116-125

Critchley M (1955) Personification of paralysed limbs in hemiplegics. Br Med J 30: 284-286

Darwin C (1877) The expression of the emotions in man and animals. John Murray, London reprinted University of Chicago Press, Chicago, 1965

Dekosky S T, Heilman K M, Bowers D, Valenstein E (1980) Recognition and discrimination of emotional faces and pictures. Brain Lang 9: 206-214

Ekman P, Friesen W V (1978) The facial action coding system. Consulting Psychologists Press, Palo Alto, Ca

Ekman P, Levenson R, Friesen W (1983) Automatic nervous system activity distinguishes among emotions. Science 221: 1208-1210

Elliott R (1969) Tonic heart rate: experiments on the effects of collative variables lead to a hypothesis about its motivational significance. J Pers Soc Psychol 12: 211-228

Gainotti G (1969) Reactions catastrophiques et manifestations d'indifference au cours des atteintes cerebrales. Neuropsychologia 7: 195-204

Gainotti G (1972) Emotional behavior and hemispheric side of the lesion. Cortex 8: 41-55

Gardner H, Lins P K, Flamm L, Silverman J (1975) Comprehension and appreciation of humourous material following brain damage. Brain 98: 399-412

Hecaen H (1962) Clinical sympomatology in right and left hemisphere lesions. In: Mountcastle (ed) Interhemispheric relations and cerebral dominance. Johns Hopkins, Baltimore

Heilman K M, Scholes R, Watson R T (1975) Auditory affective agnosia (Disturbed comprehension of affective speech). J Neurol Neurosurg Psychiatry 38: 69-72

Heilman K M, Schwartz H D, Watson R T (1978) Hypoarousal in patients with the neglect syndrome and emotional indifference. Neurology 28: 229-232

Kleck R E, Vaughan R C, Cartwright-Smith J, Vaughan, K B, Colby C Z, Lanzetta J T (1976) Effects of being observed on expressive, subjective, and physiological responses to painful stimuli. J Pers Soc Psychol 34: 1211-1218

Kolb B, Milner B (1981) Observations of spontaneous facial expression after cerebral excisions and after intracarotid injection of sodium amytal. Neuropsychologia 19: 505-514

Lacey J I (1967) Somatic response patterning and stress: some revision of activation theory. In : Appley M H, Trumbull R (eds) Psychological stress: issues in research. Appleton-Century-Crofts, New York

Lanzetta J T, Cartwright-Smith J, Kleck R E (1976) Effects of nonverbal dissimulation on emotional experience and automic arousal. J Pers Soc Psychol 33: 354-370

Lanzetta J T, Biernat J J, Kleck R E (1982) Self-focused attention, facial behavior, autonomic arousal, and the experience of emotion. Motivation Emotion 6: 49-63

Lazarus R S (1969) Patterns of adjustement and human effectiveness. McGraw-Hill, New York

Libby W L, Lacey B C, Lacey J I (1973) Pupillary and cardiac activity during visual attention. Psychophysiology 10: 270-294

Light K C, Obrist P A (1980) Cardiovascular response to stress: effects of opportunity to avoid, shock experience, and performance feedback. Psychophysiology 17: 243-252

Lubin A, Johnson L C (1972) On planning psychological research. In: Greenfield S, Sternback R A (eds) Handbook of psychophysiology. Holt, Rhinehart and Winston, New York

Malcuit G (1973) Cardiac responses in aversive situations with and without avoidance possibility. Psychophysiology 10: 295-306
Morrow L, Vrtunski B, Kim Y, Boller F (1981) Arousal responses to emotional stimuli and laterality of lesion. Neuropsychologia 19: 65-71
Obrist P A (1976) The cardiovascular-behavioral interaction - as it appears today. Psychophysiology 13: 95-107
Ross E D (1981) The aprosodias: functional-anatomic organization of the affective components of language in the right hemisphere. Arch Neurol 38: 561-569
Ross E D (1984) Right hemisphere's role in language, affective behaviour and emotion. Trends Neurosci 7: 342-346
Ross E D, Mesulam M (1979) Dominant language functions of the right hemisphere. Arch Neurol 36: 144-148
Sackeim H A, Greenberg M S, Weiman A L, Gur R L, Hungerbuhler J P, Gescwind N (1982) Hemispheric asymmetry in the expression of positive and negative emotions. Arch Neurol 39: 210-218
Sandman C (1975) Physiological responses during escape and non-escape from stress in field independent and field dependent subjects. Biol Psychol 2: 205-216
Terzian H, Cecotto C (1959) Su un nuovo metodo per la determinazione e lo studio della dominanza emisferica. G Psichiatr Neuropatol 87: 889-942
Zoccolotti P, Scabini D, Violani C (1982) Electrodermal responses in patients with unilateral brain damage. J Clin Neuropsychol 4: 143-150

Psychopathology and Hemispheric Specialization

Neural Substrates of Thought and Affective Disorders

D. M. Tucker

University of Oregon, Eugene, Oregon 97403, USA

INTRODUCTION

I will consider here two perspectives on how research on emotion and hemispheric specialization may shed light on questions of thought and affective disorders. The first perspective is a traditional one in psychological theories of emotion. It considers how cognitive processes may shape emotional experience through the labelling or elaboration of more primitive aspects of emotional arousal. This might be considered a top-down view, in which the higher cognitive capacities of the left and right cerebral hemispheres are seen as determining the nature of emotional experience.

A second perspective might be considered a bottom-up perspective. From this perspective emotional arousal may be seen not as some undifferentiated process, but rather it may have specific and unique influences on attention and memory. I will present some theoretical concepts of activation and arousal systems of the brain, and I will suggest how these may pertain to specific emotional states and specific influences on attention.
It will be the second perspective that I emphasize most in my discussion. This conference is dedicated to Giuseppe Moruzzi, who has made major contributions to our understanding of the brain's self-regulation of its levels of arousal. In remarks at this meeting, Professor Berlucchi has pointed out that in some of his later writings Moruzzi emphasized that when it was thoroughly understood he believed that the reticular activating system would be seen not as having a diffuse influence on emotion through a unidimensional effect on brain arousal, but it may have more specific influences. The theoretical model of activation and arousal systems of the brain which I will describe may offer one approach to speculating about such specific influences of brain regulatory mechanisms on emotional state. I will suggest that exaggerated levels of functioning of these neural control systems may occur in psychopathology, and that this exaggerated functioning may help explain disorders of thought and affect in psychiatric conditions.

Experimental Brain Research Series 18

Left and Right Hemisphere Function in Thought and Affective Disorders

Much of the recent research that has taken a neuropsychological approach to psychiatric conditions has examined the possibility that there may be differential dysfunction of left and right hemisphere systems in certain psychiatric conditions. Based on his observations with temporal lobe epileptics, Flor-Henry (1969) advanced a hypothesis that schizophrenia may be associated with dysfunction of the left hemisphere, whereas affective disorders may reflect greater dysfunction of right hemisphere systems. This hypothesis has generated substantial empirical and theoretical work in the last several years (for review see Flor-Henry and Gruzelier, 1983). Although there are a number of complexities shown by this research, many of the findings have been consistent with this initial hypothesis. Thus, in many schizophrenic patients, neuropsychological examination shows poor functioning of the left hemisphere, although in some conditions certain schizophrenics may show exaggerated or over-active left hemisphere functioning. Similarly, although the data on manics is mixed and controversial, there seems to be substantial evidence that severely depressed patients may have a greater impairment of right than left hemisphere cognitive skills.. The importance of considering the interaction between hemispheric dysfunction may be reversed when treatment improves the patient's mood level (Kronfol, Hamsher, Digre, and Waziri, 1978).

Thought and Affective Disorder

As Kinsbourne (this volume) has pointed out, the brain lesion evidence suggests that concepts of hemisphere functioning and emotion must be extended to consider the differential roles of anterior and posterior regions within each hemisphere. This will probably prove to be the case in research on psychiatric conditions as well. For example, in psychiatric depressives, some of the EEG evidence points to abnormalities over right anterior regions (specifically Perris and Monakhov, 1978). In research which induced a depressed affect in normal personsn, my students and I have observed that the changes in the EEG over the right hemisphere during a depressed mood showed increased activation of frontal regions (Tucker, Stenslie, Roth & Schearer, 1981). This right frontal activity was observed in the same experimental condition that produced an impairment of visual imagery, which we interpreted to reflect impairment of posterior right hemisphere functioning.

Thus, an exclusive focus on laterality may not allow a sufficient understanding of the complex interactions between anterior and posterior regions within each hemisphere.
However, at a simplistic level, the attribution of schizophrenia to a left hemisphere dysfunction and affective disorders to a right hemisphere dysfunction makes intuitive sense. There are a number of features of left and right hemisphere contributions to normal cognition and emotion that may be relevant to this general characterization.

Rationality and Emotion in the Asymmetric Brain

There have been many findings in the last decade showing the importance of the right hemisphere to normal emotional communication. Right hemisphere lesions impair the patient's ability to understand emotion communicated through tone of voice (Tucker et al. 1976) or through facial expressions (Etcoff, this volume). These findings have been supported by evidence that normal individuals show right hemisphere specialization for understanding emotional communication. Although the data on expression of emotion is not as extensive, there have been suggestions that right hemisphere lesions may impair the ability to express emotion through vocal intonation (Ross and Mesulam 1979), and the greater intensity of emotional expressions on the left side of the face in normal persons (Sackeim et al. 1978) has been taken to suggest greater right hemisphere involvement in normal emotional facial expressions.

Thus, given its importance to normal affective processes, it may not be unreasonable to suppose that the right hemisphere is dysfunctional in the psychiatric conditions described as affective disorders. Furthermore, there may be ways of reasoning from the neuropsychological research on right hemisphere cognitive skills to better understand the specific cognitive processes importance to emotional experience and behavior. In contrast to the left hemisphere's representation of cognitive processes in terms of verbal material, in which an internal code is substituted for perceptual experience, the right hemisphere seems to represent experience in an analogical code, in which the internal representation mirrors closely the individual's perceptual experience. Thus, an internal mental image of a field of flowers retains the sensory qualities of the perception, whereas a verbal representation of this experience must be an arbitrary substitution for these perceptual qualities. Could it be that the immediacy of affective processes is facilitated by this analogical quality of right hemisphere cognitive representation? Thus, as a person interprets the emotional significance of a speaker's tone of voice, the internal processing represents a direct analogue of the speaker's intonation. Would it not be the case that the listener's own emotional responses are easily engaged by this analogical representation?

Theories of emotion have often described the functions of emotions as integrating various external and internal data, including perceptual, visceral, and semantic information (Safer and Leventhal 1977; Tucker 1981). The right hemisphere's ability to integrate perceptual information in a holistic fashion may be important to this function of emotion. Drawing on the concepts of the developmental psychologist Heinz Werner, I have suggested that the right hemisphere may function in a syncretic mode, in which various aspects of experience are apprehended in a holistic, undifferentiated fashion. In a recent theoretical formulation of emotion, Buck (1985) proposes that syncretic cognition in the right hemisphere may be a mechanism through which emotional experiences become available to the cognitive system.

Thus, it may be that neuropsychological research and theory may suggest ways of understanding the specific information processing operations that allow the brain to handle emotional information. From this approach, emotions will be seen not just as subjective phenomena, but as specific mental operations that have adaptive significance.

Although the research on left hemisphere cognitive processes has tended not to emphasize emotional implications, and it often seems implied that the left hemisphere is simply non-emotional, there are characteristics of left hemisphere cognition that may be quite relevant to a general theory of emotion and the mind. Many of the mechanisms of rational thought would seem to be supported by the specific cognitive characteristics of the left hemisphere. Thus, the left hemisphere's representation is in a substitutive mode, where words stand for ideas. The left hemisphere formulates ideas in a propositional fashion, framing active statements about the nature of reality. This propositional formulation seems to be essential for logical reasoning.

In addition, several research findings suggest that the left hemisphere is particularly important to sequential control, both of cognition and motor behavior (Cohen 1973). This kind of sequential control may be essential for logical reasoning, in which the relation between one proposition and the next can be explicated, and in which the order between propositions can be reversed to examine the integrity of the causal linkage. Although it may be argued that we do not have sufficient basis for attributing rationality to one half of the brain, it is very difficult to consider how these specific requirements for rationality could be mediated by right hemisphere systems.
Thus, we have a model of cognitive characteristics of the left and right hemispheres that may explain some aspects of the hemispheres' roles in general mental functioning, and in emotion in particular. Dictionary definitions of rationality may not characterize the formal mental properties that support rational thought, but they often state that rational thought is opposed to emotionality. Thus, it may be that the more sequential and analytic ideation of the left hemisphere is important in providing a cognitive domain that is relatively uninfluenced by affective processes.

Limits on the Rational-Emotive Dichotomy

Although I think these general characteristics of left and right hemisphere functioning are essentially correct, they are also incomplete. There are important aspects of coherent thought even in the verbal domain that seems to be supported by right hemisphere cognition. In addition, there may be suggestions that the left hemisphere is important to emotional stability.

Studies of story comprehension both with and without humor, have suggested that the intact right hemisphere may be required for integrating the various facts in the story into a coherent understanding (Gardner et al. 1975). Thus, the patient with an intact left hemisphere but a damaged right hemisphere may be able to understand each of the verbal messages of the story, but may be deficient in relating various semantic elements and constructing a coherent story theme. Even with linguistic material, then, the right hemisphere appears to integrate information in a way that may be essential for rational understanding.

The left hemisphere's functioning may not be irrelevant to affective experience. Some evidence has suggested that the left hemisphere may normally inhibit and regulate right hemisphere affective responsiveness (Buck and Duffy 1980; Tucker and Newman 1981). In addition, studies of individuals who are characteristically anxious have suggested that they may show exaggerated left hemisphere functioning in perception and attention (Tucker et al. 1978; Tyler and Tucker 1982). Highly anxious subjects may show an overly analytic perceptual mode, suggesting exaggerated left hemisphere functioning.

Whereas the information in the right hemisphere's emotional processes seems easy to relate to the concept of right hemisphere dysfunction in affective disorders, the evidence on the left hemisphere's role is somewhat more quizzical. Thus, left hemisphere dysfunction in schizophrenia seems relevant to thought disorders, but there may be exaggerated left hemisphere functioning in this disorder. It is not the case that the schizophrenic simply lacks verbal ability, but rather he shows disorganized thought. One hypothesis that my colleagues and I have entertained is that there may be a parallel between the exaggerated left hemisphere functioning in anxious normals and the left hemisphere overactivation in schizophrenia (Tucker et al. 1978).

Asymmetry of Activation and Arousal Systems

Another perspective on abnormal hemispheric function in psychiatric disorders can be gained from the bottom-up view, in which asymmetries of elementary regulatory mechanisms may explain important aspects of hemispheric emotional and attentional functioning. For several of the primary neurotransmitter pathways that are implicated as dysfunctional in psychiatric disorders, recent findings have suggested asymmetric distributions between the left and right hemispheres (reviewed by Tucker and Williamson 1984). If it were possible to understand the functional significance of these asymmetries, it may be that we can reason from the biological approaches to psychiatric disorders, which emphasize neurochemical alterations, to the neuropsychological approaches, which emphasize lateralized hemispheric dysfunction. Perhaps more importantly, the neurotransmitter pathways are the specific mechanisms through which the brain self-regulates its level of alertness and arousal. Thus, it may be possible to begin to characterize the ways that specific components of the reticular activating system modulate attentional competence and emotional arousal.

Perhaps the most important principle to be gained from the literature on neurotransmitter pathways is that alterations of the functioning of these systems is not a simple quantitative increase in brain activity, but rather a qualitative change in neural functioning (Tucker and Williamson 1984).

Drawing on Pribram and McGuinness' (1975) model of attentional control, Williamson and I differentiated between a tonic activation system that supports motor readiness and a phasic arousal system that supports perceptual responsivity. Each of these systems appears to have a unique set of neurotransmitter mechanisms, and has qualitative influences on attention.

The dopamine pathways seem to be integral to motor readiness. Consistent with this role, pharmacologic stimulation of these pathways increases motor activity. However, it is not the case that continued increases in dopaminergic modulation produce simple quantitative increases in motor functioning. Rather, what occurs is a qualitative change in motor control. With high levels of dopaminergic modulation, the animals begin exhibiting routinized, repetitive motor acts (Iversen 1977). With excessive levels, they show stereotyped motor sequences. This is an interesting control mechanism. At the same time as its mechanism augments motor functioning, this mechanism restricts the range of actions that are engaged in. Williams and I characterized this as a redundancy bias that appears integral to the tonic activation system.

In contrast, the phasic arousal system shows a habituation bias. The norepinephrine pathways that appear integral to this system are essential to alerting and wakefulness, yet they are also important to habituation and behavioral extinction. Thus, the control system that supports perceptual orienting also seems to apply a qualitative influence on perceptual responsivity, such that sameness of perceptual surroundings decrements the brain's response to perceptual events. The positive characteristic of this control mechanism is an orienting to novelty in the environment.

Recognizing the qualitative influences of these control systems allows some interesting theoretical specualation on how neurophysiologic mechanisms can regulate the scope and terminal stability of attention. The tonic activation system provides stability and continuity to motor control; it may also provide similar regulatory influences for higher order cognitive processes. The phasic perceptual arousal system has integral control features that allow for the selection for novelty; this system may also support not just elementary but more cognitive aspects of the brain's perceptual operations.

These regulatory systems have been implicated not only in activation and arousal, but in emotional states. The dopamine pathways have been implicated in schizophrenia. Williamson and I proposed that the control of motor readiness may be important to not just the cognitive aspects of vigilant attention, but the subjective aspects of anxiety. Thus, at the attentional level, the motor readiness system engages an active and vigilant focusing of attention; at the phenomenological level, this is experienced as anticipation of possible threatening events.
The norepinephrine pathways have been implicated in affective disorders as suggested by the catecholamine hypothesis (Schildkraut 1965). Williamson and I proposed that the affective dimension of depression-elation may be the subjective experience that emerges from variations in the functioning of the phasic arousal system. If these formulations are correct, there may be an

inherent relation between subjective affective states and attentional capacity.

Activation, Arousal, and Hemispheric Cognitive Functioning

What is the functional significance of the apparent asymmetries in the distribution of the neurotransmitter pathways supporting the activation and arousal systems? Williamson and I proposed that the asymmetries can be understood by appreciating the different roles of the tonic activation and phasic arousal systems in motor and perceptual operations, respectively, and by appreciating the differential specialization of left and right hemispheres for motor and perceptual operations, respectively.

The neurological literature has shown the importance of left hemisphere function for fine motor and sequential motor control (Heilman 1979). We have suggested that in its specialization for fine motor control, the left hemisphere has come to rely on the attentional characteristics of the tonic motor readiness system. Indeed, the left hemisphere's skills in sequential processing and analytic cognition may be seen as consistent with the specific control effects of the tonic activation system. Consider the influence of the redundancy bias on perceptual operations. This bias causes a few perceptual elements to be retained in working memory, and it causes these elements to have substantial continuity across time. In a spatial frame, this would result in attention being focused on one or a few aspects of the perceptual array, rather than being allocated evenly across the entire perceptual array. Thus, a focusing of attention can be seen as consistent with the elementary control of a redundancy bias. This attentional focusing may be essential to allow the left hemisphere its skills in analytic cognition.
It may also be the case that the right hemisphere's skill in holistic cognition can be seen as consistent with the regulatory influences of the phasic arousal system. Consider the effects on working memory of novelty selection bias of the phasic arousal system. By orienting perception toward many diverse novel elements, this mechanism would fill short-term memory with a wide array of unique data. This would produce an expansive attentional scope in which many elements were contained simultaneously in working memory. Would not this expansive attentional scope support the right hemisphere's holistic perception and cognition?

Thought and Affective Disorders Resulting From Excessive Activation and Arousal

Formulated in this way, this theoretical model makes specific predictions about the influence of anxiety and depression-elation on attentional and cognitive processes that may be relevant to the observations of disordered mental functioning in psychopathology. There are several observations on the effects of anxiety on cognition that are consistent with the qualitative cybernetics of the tonic activation system. For example, anxiety has been found to be associated with a decreased utilization of peripheral cues in attentional experiments (Easterbrook 1959). This may be consistent with the

excessive focusing of attention and overly analytic perceptual mode that would result from a high degree of functioning of the tonic activation system with high anxiety. Anxiety has also been observed to impair memory for remote associates in a verbal learning task (Spence 1958). Although this has been explained as a facilitation of dominant responses in the habit family hierarchy, it would also be consistent with a redundancy bias which restricts novelty in associational as well as attentional function. At more extreme levels, the tonic activation system would produce stereotypic, routinized behavior both in cognition and motor actions. This seems directly relevant to the symptomatology of obsessive-compulsive disorders. At greater levels of disorganization, this stereotype might be seen in the disorganized motor behavior and repetitive covert verbalizations of the schizophrenic.
An opposite influence on attentional scope would be predicted for relatively lower levels of anxiety and higher levels of elation. Thus, it would be predicted that manics would show an overinclusive associational process, and this is what has been observed (Andreasen and Powers 1975). This over-inclusiveness in mania may not only reflect a kind of thought disorder in pathologic brain, but may represent the extreme form of an attentional mode that may be adaptive at less pathological levels of elation. Creative artists have a higher than normal incidence of affective disorder (Andreasen and Powers 1975). It is often the case that manic patients complain that their creativity suffers when they are treated with lithium. Isen and associates (Isen 1984) have found that inducing a more elated affect in normal individuals can also produce a more expansive attention. Experimentally inducing positive affect was found to result in an intuitive rather than methodical approach to problem solving. Positive affect also resulted in greater novelty in responses to tasks requiring creative problem solving. Finally, isen and Daubman (1984) observed that experimental induction of positive affect in normals resulted in more remote associates to a word association task. Thus, normal elation may not only improve creativity, but it may produce structural changes in attention that parallel closely those which become pathological in mania.

Conclusions

In this discussion I have considered disordered mental functioning in psychiatric conditions from two perspectives, one which emphasizes hemispheric function at the cognitive level and another which emphasizes lateral asymmetries of primitive control mechanisms. From both these perspectives, the disordered mentation seen in psychiatric conditions is not simply a pathological condition, but rather it must be understood as an extreme form of neural mechanisms which are under most conditions adaptive. Thus, to understand the apparent disorders of hemisphric function in psychopathology, we require a theory of hemispheric specialization for normal emotion.

It is also the case that from the perspectives I have considered, we find a very close relation between emotional control mechanisms and the attentional capacities of the left and right hemispheres. Thus, not only do extreme emotional states disorganize attention, but optimal levels of these same emotional states would be required fr optimal cognitive functioning. This suggests an interesting way of understanding emotion that is perhaps novel for psychological theories of emotion. From this way of thinking, emotions are not simply disorganizing influences on mentation, nor are they simply subjective phenomena. Rather they exert specific and essential regulatory controls on attentional functioning and cognitive organization. Emotions may indeed be the engines of the mind.

REFERENCES

Andreasen N J C, Powers P S (1975) Creativity and psychosis: an examination of conceptual style. Arch Gen Psychiatry 32: 70-73

Buck R (1985) Prime theory: an integrated view of motivation and emotion. Psychol Rev 92: 389-413

Buck R, Duffy J (1980) Nonverbal communication of affect in brain-damaged patients. Cortex 16: 351-362

Cohen G (1973) Hemispheric differences in serial versus parallel processing. J Exp Psychol (Gen) 97: 349-356

Easterbrook J A (1959) The effect of emotion on cue utilization and the organization of behavior. Psychol Rev 66: 183-201

Flor-Henry P (1969) Psychosis and temporal lobe epilepsy: a controlled investigation. Epilepsia 10: 363-395

Flor-Henry P, Gruzelier J (eds) (1983) Laterality and psychopathology vol 2. Elsevier, Amsterdam

Gardner H, Ling P K, Flamm L, Silverman J (1975) Comprehension and appreciation of humorous material following brain damage. Brain 98: 399-412

Heilman K M (1979) Apraxia. In: Heilman M, Valenstein E Clinical Neuropsychology. Oxford University Press, New York

Isen A M (1984) Toward understanding the role of affect in cognition. In: Jr. Wyer R S, Srull T K (eds) Handbook of social cognition, vol 3. Lawrence Erlbaum, Hillsdale, NJ

Isen A M, Daubman K A (1984) The influence of affect on categorization. J Pers Soc Psychol 47: 1206-1217

Iversen S D (1977) Brain dopamine systems and behavior. In: Iversen L L, Iversen S D, Snyder S H (eds) Drugs, neurotransmitters and behavior. Plenum: New York, pp 333-384 (Handbook of psychopharmacology, vol 8)

Kronfol Z, Hamsher K, Digre K, Waziri R (1978) Depression and hemisphere functions: changes associated with unilateral ECT. Br J Psychiatry 132: 560-567

Perris C, Monakhov K, VonKnorring L, Botskarev V, Bikiforov A (1978) Systemic structural analysis of the EEG of depressed patients. Neuropsycholbio 4: 207-228

Pribram K H, McGuinness D (1975) Arousal, activation, and effort in the control of attention. Psych Rev 82, 2: 116-149

Ross E, Mesulam M M (1979) Dominant language functions of the right hemisphere? Prosody and emotion gesturing. Arch Neurol 36: 144-148
Sackeim H A, Gur R C, Saucy M C (1978) Emotions are expressed more intensely on the left side of the face. Science 202: 434-436
Safer M A, Leventhal H (1977) Ear differences in evaluating emotional tone of voice and verbal content. J Exp Psychol (Human Percept) 3: 75-82
Schildkraut J (1965) The catecholamine hypothesis of affective disorders: a review of supporting evidence. Am J Psychiatry 122: 509-522
Spence K W (1958) A theory of emotionally based drive (D) and its relation to performance in simple learning situations. Am Psychol 13: 131-141
Tucker D M, Watson R G, Heilman K M (1976) Affective discrimination and evocation in patients with right parietal disease. Neurology 26: 354
Tucker D M, Antes J R, Stensile C E, Bernhardt T N (1978) Anxiety and lateral cerebral function. J Abnorm Psychol 87: 380-383
Tucker D M (1981) Lateral brain function, emotion and concept-utilization. Psychol Bull 89: 19-46
Tucker D M, Newman J P (1981) Verbal versus imaginal cognitive strategies in the inhibition of emotional arousal. Cogn Ther 5: 197-202
Tucker D M, Williamson P A (1984) Asymmetric neural control systems in human self-regulation. Psychol Rev 91: 185-215
Tyler S K, Tucker D M (1982) Anxiety and perceptual structure: individual differences in neuropsychological function. J Abnorm Psychol 91: 210-220

Reaction Times and Mental States

G. Rizzolatti

Institute of Human Physiology, University of Parma, via Gramsci 14, 43100 Parma, Italy

INTRODUCTION

The notion that the time necessary for mental operations can be measured was introduced by Donders (1869) at the end of the last century. Donders measured the time employed by normal subjects to respond to sensory stimuli. He found that when the task required a stimulus discrimination or a response choice, the reaction time was longer than when only one stimulus was presented and a predetermined response was required. He concluded that the increase in reaction time observed in choice situations measured the time necessary to accomplish the mental operations involved in stimulus or response selection. At the beginning of the century Poffenberger (1912) employed the technique developed by Donders to test separately the two cerebral hemispheres. Poffenberger presented visual stimuli to the right or to the left of a fixation point and instructed the subjects to press a button as fast as possible with the right hand or with the left hand. His results showed that the responses to right stimuli were faster with the right hand than with the left hand, whereas those to left stimuli were faster with the left hand than with the right one. Poffenberger's interpretation of his findings was that the uncrossed responses (stimulus and effector on the same side) were faster than crossed responses (stimulus and effector on the opposite sides) because the former could be mediated by one hemisphere, whereas the latter require the activity of both hemispheres.

The reaction time method, after a long period of eclipse, has been rediscovered in modern times and, associated with the lateralized stimulus presentation first adopted by Poffenberger, is now one of the basic tools for studying hemispheric specialization in normal subjects. The most commonly used experimental situation is the following. Complex stimuli (e.g., words, faces) are presented to the right or to the left of the fixation point. The subjects are instructed to respond to some of them and not to others. The discrimination between stimuli is necessary in order to have responses dependent upon the specific content of the stimuli and not upon the stimulus' appearance. Reaction times of the right hand and of the left hand are measured in response to right and left stimuli. The comparison between the four stimulus-response conditions allows one to decide whether one of the two hemispheres is superior to the other in processing specific stimuli.

Experimental Brain Research Series 18

The method of measuring reaction times to complex stimuli has some important limitations. The major one is that only the perceptual aspect of the hemispheric organization is tested. No information can be obtained on the way in which cortex is involved in movement control. Furthermore, this method does not appear to be suitable for studying processes internally generated by the subjects or the emotions. Prompted by these facts, some years ago we devised a technique, based on simple reaction times, which tests independently the two hemispheres during cognitive tasks and emotions. In this article I will review a series of data obtained with this technique.

INTERFERENCE TECHNIQUE

In the absence of a cognitive task and mood alterations, the response speed of the two hemispheres to unstructured visual stimuli is identical (see Rizzolatti 1979; Bashore 1981). In some experiments an advantage of the responses mediated by the right hemisphere has been reported, but even in these cases the interhemispheric difference was small and inconstant (Jeeves and Dixon 1970; Anzola et al. 1977; Berlucchi et al. 1977). The idea at the basis of the interference technique is that this fundamental hemispheric symmetry is altered when a task requires the specific abilities of one hemisphere. In this condition the hemisphere involved in the task should have difficulties in executing a second operation, even so simple as that of responding to a flash of light.

In an initial series of experiments (Rizzolatti et al. 1979) we tested this hypothesis by employing tasks proper to the left hemisphere. The prediction was that an increase of reaction times mediated by the left hemisphere would occur. The experiments were carried out on normal, right-handed male subjects. The basic experimental situation was the following. The subject sat in a soundproof room facing a translucent, uniformly illuminated screen. The visual stimulus was a circle of light having a diameter of about 1 degree. It was projected on the screen 5 degrees to the right or to the left of the fixation point. The subject was instructed to maintain central fixation and his eye position was continuously monitored by close circuit T.V.

An acoustic "ready" signal preceded the light by a variable interval of 1-3 s. At the light stimulus presentation, the subject had to press a microswitch with the thumb of one hand as quickly as possible. There were four blocked conditions: stimulus to the right, response with the right hand; stimulus to the right, response with the left hand; stimulus to the left, response with the left hand; stimulus to the left, response with the right hand.

The first cognitive task we studied was a praxic task. The subject had to perform a predetermined, sequential movement of single fingers either with his right or with his left hand. The following sequence was used: thumb, middle finger, index finger, ring finger. The sequence was repeated without stopping. While performing the task with one hand, the subject had to respond to light stimuli with the other. The results showed that the introduction of the praxic task produced three effects: (a) a global increase of reaction times; (b) a selective lengthening of the responses of the right hand; (c) a selective lengthening of the responses mediated by the left hemisphere. The inferiority of left hemisphere in mediating light responses was found in all subjects tested (n = 8). The difference between left hemisphere and right hemisphere responses ranged from 8 ms to 26 ms (average 16 ms).

Similar results were obtained when the subjects performed a verbomotor task. In this experiment they were asked to count backwards by 3s or 4s from a starting point indicated by the experimenter. The subject continued the subtracting until reaching zero at which time the experimenter redefined the starting point and subtracting factor. The presentation of the light stimulus was carried out as in the previous experiment. The results showed a global increase of the reaction times and a specific lengthening of those mediated by the left hemisphere. No significant differences were found between hands.

We interpreted the global effect and the selective hemispheric difference as follows. For the global effect we suggested a "capacity interference" (Kahneman 1973). Each subject has a limited amount of attention and while engaged in programming and executing the praxic or verbomotor task he allocates it to these tasks. As a consequence he cannot respond efficiently to the incoming flashes of light. A similar phenomenon should be also responsible for the lengthening of the right hand responses observed in the praxic task. As mentioned above, the experiments were carried out on right-handed subjects, on subjects therefore who had more difficulties in performing the task with the left hand than with the right hand. For this reason they had to allocate more attention to the praxic task when it was executed with the left hand than when it was executed with the right hand. As a consequence, the right hand responses were interfered with more than the left hand responses. This interpretation accounts for the lack of differences between hands in the verbomotor task. Since in this task the subjects' attention was equally engaged in the case of right hand or left hand responses the difference between hands disappeared.

The most interesting result of the two experiments was the selective increase of reaction times mediated by the left hemisphere. Since the tasks selected for the experiments are known to require the specific abilities of the left hemisphere, we concluded that the activation of this hemisphere was responsible for the effect. The presence of a selective hemispheric interference can be explained in two ways. The first is that we were dealing with a particular case of capacity interference. The second is that a structural interference was responsible for the effect.

Capacity Versus Structural Interference

According to the capacity interference hypothesis, each hemisphere has its own fixed amount of attention. When one hemisphere is specifically engaged in a task, its attention becomes insufficient for other tasks. Hence, the hemisphere which is not engaged in the task gives fast responses, the engaged one gives slow responses. According to the structural interference hypothesis, the delay of the engaged hemisphere is due to a competition between the two tasks for the same neural structures. In particular, in the two experiments described above the organization of finger or speech movements impairs the responsiveness of the left hemisphere.

The two hypotheses can be tested by asking the subjects to perform lateralized tasks that require considerable attention, but are free from any obvious motor organization aspects. The capacity interference hypothesis predicts a lengthening of the responses of the hemisphere specifically engaged in the task even if the stimuli are given in different sensory modalities. In contrast, the structural interference hypothesis predicts an absence of interference.

In order to test these hypotheses we presented to a group of subjects strings of four digits, alternated with strings of three digits and one letter (Rizzolatti et al. 1982). The various strings were recorded on a magnetic tape and presented to the subjects by means of a loudspeaker positioned above their head. The letters were all bisyllabic and chosen from Latin and Greek alphabets (Italian pronunciation). The letter could occur in the place of the first, second, third, or fourth digit. The subjects were required to detect the letter and to repeat it as they heard it. The light stimulus for reaction times was randomly presented after one of the four digits of the string.

The results showed a global increase of the reaction times, but no hemispheric differences. It is clear therefore that to pay attention to certain types of stimuli is not sufficient to increase selectively the reaction times of the hemisphere selectively engaged in processing these stimuli.

This conclusion was confirmed by a second experiment in which again strings of digits were presented to the subjects. Unlike in the previous experiment all strings were formed by digits. The subjects' task was to listen to the digits and to repeat them aloud at the end of each digit string in the same order as they had been presented (memory task). The results showed a global increase of reaction times, without any impairment of those mediated by the left hemisphere.

Finally, in a third experiment the same strings of digits were again presented but, this time, the subjects' task was to repeat each digit aloud as soon as they heard it. It is well known that this task ("shadowing") requires little attention to be executed. It requires however a certain amount of motor organization. The results showed a slight lengthening of reaction times mediated by the left hemisphere. Thus, in the absence of strong attentional demand there was no global increase of reaction times, but the necessity of motor programming produced an impairment of the left hemisphere responsiveness.

Once established that the increase of the left hemisphere reaction times during praxic and verbal tasks is due to a structural interference, the problem arises of the locus where this interference may occur. In theory, interference may take place at the stage where the stimulus is analyzed, at the stage where motor plans are elaborated, or at the stage where the motor plans are implemented. Of course, these possibilities are not mutually exclusive and a combination of factors could be responsible for the effect. However, the experimental data indicate that the interference occurs at a processing level preceding that of motor execution. If the interference had been at this level, in the counting backwards experiment the right hand should have been consistently slower than the left hand regardless of the visual field of presentation. In contrast, no significant difference between hands could be seen.

There is insufficient evidence to decide whether the cognitive tasks interfere with stimulus processing already at the sensory level or at higher, "premotor" levels. The shortest anatomical pathway that leads visual information to cortical motor centers responsible for the responses is made up of the primary visual area, the constellation of visual areas located in the peristriate region, the posterior parietal lobe, and the premotor areas of the frontal lobe. A rigid structural interpretation of the interference effect postulates that this effect takes place as a consequence of a competition between the various signals for neural substrates. For example, in the case of the praxic task the command "move the thumb", triggered by the visual stimulus, cannot be immediately executed because the cortical area responsible for the organization of this movement is engaged in the organization of sequential finger movements. This competition within a same nervous center produces the slowing of the response. A less rigid interpretation of the interference effect postulates that the motor activity produces, in parallel with the activation of the circuits responsible of the motor program, an inhibition of those circuits whose activation would determine other types of motor programs. This idea is congruent with well-established data on movement control showing that excitatory and inhibitory circuits are simultaneously activated in normal conditions. For example, a reflex activation of spinal cord extensor motoneurons is accompanied by a simultaneous inhibition of flexor motoneurons and, vice versa, an excitation of flexors is accompanied by an inhibition of extensors. This mechanism of reciprocal inhibition (Sherrington 1906) allows the execution of reflex movements without obstacle from antagonist muscles. Reciprocal inhibition intervenes not only in the case of elementary movements such as flexion-extension, but also when two reflexes produce contrasting effects. In the dog, stimulation of the back evokes the well-known scratching reflex (Sherrington 1906). The scratching movements are executed by the hindlimb ipsilateral to the stimulated side. Simultaneously, the contralateral hindlimb extends in order to provide a support for the animal. If the skin of the animal is stimulated bilaterally, the reflex appears on one side only. Two scratching reflexes are incompatible, since their simultaneous occurrence would prevent the upright posture. Thus, one inhibits the other. A structural interpretation of the interference effect which does not postulate a competition for neural substrate in a same center appears to be more plausible than the rigid version of it. Its greater explanatory value will be apparent after discussion of data on emotions.

Right Hemisphere Interference

Up to now all examples of hemispheric interference I have illustrated concerned the left hemisphere. This might suggest that the right hemisphere is able to elaborate information in parallel and organize simultaneously several activities. Experiments where the hemispheric responsiveness was tested during the programming of eye movements prove that it is not true (Sava et al. 1988).

The experimental situation was the following. The subject sat in front of a screen on which different types of mazes were projected. Each maze had four entrances and one exit. The subject was instructed to explore the mazes, starting from the first entrance, and to indicate which path leads to the exit. Both eyes were open and the subject was allowed to move them in order to solve the problem. During the exploration a visual stimulus was presented 20 degrees to the right or the left of the maze center. The length of the maze was 8 degrees. As in the previous experiments, the subject had to respond as fast as possible, pressing a switch at the appearance of the visual stimulus. After some practice all subjects were able to solve the problem. Since the maze exploration is a discontinuous task, in which eye movements alternate with periods of rest, some light stimuli necessarily fell in these rest periods. In order to have more homogeneous data, for each subject two reaction-time distribution histograms were constructed. The first one consisted of reaction times during the oculomotor task, the second one of reaction times in the absence of any interfering activity. For the statistical analysis those reaction times were used which, in the histogram with the oculomotor activity, fell above a value established on the basis of the histogram without the oculomotor activity.

The results showed that, regardless of the employed hand, the reaction times mediated by the left hemisphere were significantly faster than those mediated by the right hemisphere. The selective impairment of the right hemisphere responses was present in 11 out of 12 subjects. No other factor or interaction was significant at the statistical analysis.

These findings were confirmed by another experiment (Sava et al. 1988) in which the subjects were required to follow with their eyes a spot of light moved at a variable speed and with variable trajectories on the screen of a computer. Also in this situation the right hemisphere reaction times were lengthened. The effect was present in 10 out of the 12 subjects tested. The intensity of the interference ranged from 4 to 37 ms, with an average of 13 ms.

It is clear from these data that the interference produced by the motor programming is not a phenomenon limited to the left hemisphere. Oculomotor tasks give results similar to those obtained with verbomotor or praxic tasks but the interference concerns the right hemisphere. The difference is due to the different hemispheric localization of the circuits responsible for the praxic and verbomotor tasks on one side, and oculomotor tasks on the other. On the contrary, there is no evidence that the right hemisphere can organize simultaneously several tasks.

INTERFERENCE TECHNIQUE AND EMOTIONAL STATES

An interesting problem which may be addressed with the interference technique is that of emotional states. There is little doubt that every emotional state generates a series of motor and vegetative responses. One should expect therefore that if one hemisphere is predominantly responsible for a given emotional state and its behavioral counterpart, its response to a test stimulus should be lengthened, as it occurs during hemisphere-specific cognitive activities. The interest in a study testing this hypothesis is increased by the fact that, whereas there is vast agreement that the recognition of emotions is chiefly a function of the right hemisphere (Ley and Bryden 1979; Tucker 1981; Bear 1983), the role of the two hemispheres in the production of emotions remains largely unknown (see Gainotti 1972 and also the chapters by Caltagirone, Gainotti and Pizzamiglio, this volume).

Reaction times during emotions were studied in a group of female subjects (Ladavas et al. 1984). Each subject was tested in three experimental sessions on three different days. In each session there were two conditions. In the first condition the subjects had no other instruction except that of responding to the visual stimulus, in the second condition they had to respond to the same stimulus but after induction of a negative affect. The induction of negative affect was made by simply asking the subject to feel sad. The instruction was: "You must feel sad; it is not difficult, try it. Do not imagine unhappy scenes. It is useless. Just think that you are sad, very sad. When you are ready we will start the experiment." Although it may seem surprising, all subjects succeeded in self-inducing the emotion.

The results are shown in Fig. 1. It is clear that during the emotional state there was a marked global lengthening of the reaction times. Moreover, whereas in the control condition there was no difference between the two hemifields, during the emotional state the responses mediated by the right hemisphere were longer than those mediated by the left hemisphere. An analysis of variance showed that the factor presence-absence of emotion and the interaction between this factor and visual hemifield were significant. The disadvantage of the right hemisphere was present in all subjects but one (n = 12). This disadvantage ranged from 3 to 23 ms.

One objection that may be raised against this experiment is that the subjects, although expressly asked not to imagine visual scenes, did not comply to the

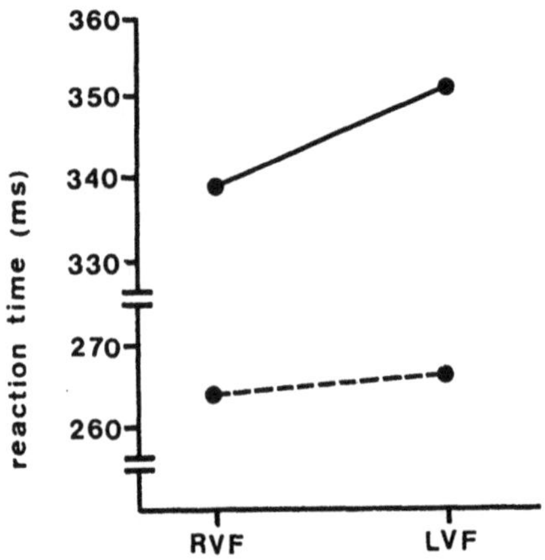

Fig.1 Reaction time and sadness. Mean reaction times during test session (continuous line) and control sessions (interrupted line). RVF, right visual field; LVF, left visual field

instruction and used this last strategy in order to induce the negative affect. If this was so, one may argue that the lengthening of the right hemisphere reaction time was due to the process of imaging and not to the induced emotion. In order to control this possibility, a second experiment was carried out having the same experimental design, but another interfering task. This time the subjects, again all women, were required to imagine themselves in three daily-life situations, devoted of emotional content. Three situations, one for each session, were chosen. In the first, the subjects had to visualize their breakfast, in the second their morning toilet, in the third their dressing. The instruction was to recollect visually, as meticulously as possible, the room where the different scenes took place, the objects they used, and their own acts. All subjects told the experimenters that they were able to perform the task.

The results are shown in Fig. 2. From the strong increase of the reaction times, one can see that the subjects had their attention engaged in the task. An analysis of variance demonstrated that this increase was significant. No hemispheric differences, however, were found. This result rules out the possibility that the hemispheric differences found during negative affect were due to visual imagery; in contrast, it supports the hypothesis that an emotional state may modify the hemispheric responsiveness.

Unlike the experiments in which a negative emotion was induced, experiments with positive emotions gave unsatisfactory results. Most subjects had difficulties in self-inducing a positive affect as well as in maintaining it for the whole experimental session. The results failed to show any hemispheric difference. The validity of these results is obviously dubious. It is possible that reliable results could be obtained by using more sophisticated induction

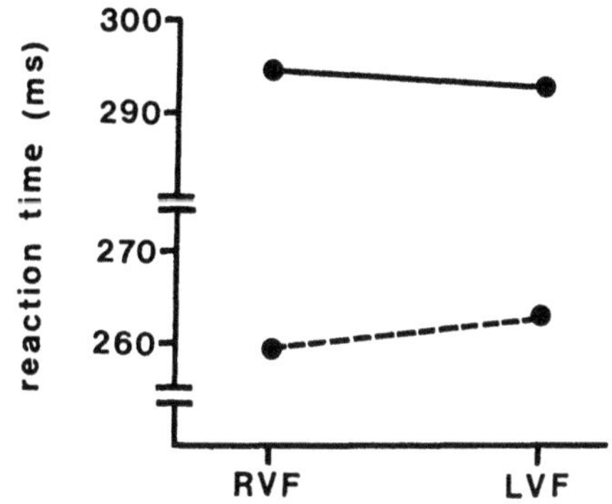

Fig.2 Reaction times and visual imagery. Mean reaction times during test sessions (continuous line) and control sessions (interrupted line). Abbreviations as in Fig.1

techniques such as those described by Ekman et al. (1983) and Bloch et al. (1987).

Reaction Time and Mood Disturbances

The finding that during negative affect a mechanism is activated which interferes with the right hemisphere responsiveness suggests that a similar mechanism may be hyperactive in patients with mood disturbances. In order to test this hypothesis an experiment was carried out on a series of patients subdivided into two experimental groups and one control group (Liotti et al. 1988). The first experimental group consisted of eight outpatients with a diagnosis of unipolar nonpsychotic depression. The severity of the depression during the experimental period was measured using the test of Hamilton (1960). The average score was 37.9. All patients included in this group had very few signs of anxiety.

The second group was formed by 12 patients for which anxiety was the major disturbance. Only patients with peak values at the Cattell scale of anxiety (9-10) were accepted in this group. The absence of severe depressive symptoms was established by analyzing the values of the subscore "depression" on the Hamilton test. The average of this value was 6.3. This value is significantly different from that of the other experimental group (8.8). Finally, dermatological patients, matched for age and social condition to the psychiatric patients, were used as controls.

The experiments were carried out in the same way as the experiments on normals, except for the lack of an interfering task. The tested hypothesis was that the mood alteration impairs the right hemisphere responsiveness. The results confirmed it. In the depressed patients the reaction times mediated by the right hemisphere were significantly longer than those mediated by the left hemisphere. This increase was present in all patients, with a range between 9 and 17 ms (average, 11 ms).

The group of patients affected by anxiety showed a slight but significant increase of reaction times in comparison with the controls. This increase was due entirely to a lengthening of left hemisphere responses. The difference, however, between reaction times mediated by the right and left hemispheres was not significant. The discrepancy between the statistical analyses renders difficult the interpretation of these data. Finally, the control group did not show any significant difference between responses in various conditions.

How can one interpret these data and, in particular, the increase of reaction times mediated by the right hemisphere of the depressed patients? There seem to be two most likely interpretations: the first is that depressed patients have a lesion or a functional impairment of the right hemisphere. The second is that in depressed patients there is a hyperactivity of circuits responsible for negative mood and that these circuits are located in the right hemisphere. This hyperactivity might produce as a secondary fact an inhibition of the circuits involved in reaction times and possibly of other types of responses mediated by the right hemisphere.

There is evidence that suggests that the hyperactivity hypothesis is the most likely. First, this hypothesis is congruent with the findings in normal subjects in which a negative mood has been induced (Ladavas et al. 1984). To postulate opposite mechanisms for negative mood in normals and in depressed patients is not parsimonious.

A second argument in favor of the hyperactivity hypothesis derives from electrophysiological studies in patients with mood disorders. Although, as it may be expected, the EEG pattern of these patients is multifarious and variable, nevertheless, the emerging picture is that depression is characterized by an activation of the right hemisphere and in particular of the right frontal region. The EEG asymmetry is especially clear in patients without or with modest signs of anxiety (Perris and Monakov 1979; Perris et al. 1979).

In favor of right hemisphere hyperactivity during negative mood are also EEG data obtained in normal subjects. Davidson et al. (1979) have shown that during artificially induced negative affect, the right frontal cortex shows a greater desynchronization than the left frontal cortex. Conversely, during positive affect an increase in desynchronization is observed in the left frontal region. A frontal lobe EEG asymmetry due to mood alterations was reported also by Tucker et al. (1981). Also these authors observed an increase of activity in the right frontal region during artificially induced negative mood.

In apparent contrast with the hyperactivation hypothesis are the findings that depressed patients frequently show an impairment in executing visuo-spatial tasks (Kronfol et al. 1978; Fromm-Auch 1982). As it is well known these tasks are a function of the right hemisphere. It appears therefore that in depressed patients there is a functional or organic lesion of the right hemisphere. The considerations on structural interference discussed above allow one, however, to explain the right hemisphere impairment without postulating a lesion of this hemisphere. If during depression, as during sadness, there is a hyperactivation of the circuits which mediate the negative affect, it follows from the law of reciprocal inhibition that this hyperactivity should determine an inhibition, at least partial, of other activities mediated by the right hemisphere. The deficit in the visuospatial tasks will be a consequence of this inhibition.

Congruent with this interpretation is the finding that electroconvulsive therapy, especially when performed unilaterally on the right side, produces in addition to an improvement of the mood a recovery of the visuospatial capabilities (Kronfol et al. 1978; Fromm-Auch 1982). This improvement should occur because the therapy interrupts the excessive activity of the right hemisphere emotional circuit. In this way it eliminates the source of inhibition and restores the normal functioning of the circuits responsible for the visuospatial abilities. By contrast, it is hard to understand how an electroconvulsive therapy could improve a real organic damage.

In conclusion, the interference technique is a technique easy to handle and which gives information on the involvement of the two hemispheres in cognitive tasks and in emotional states. Its use for studying emotions appears to be particularly promising since new powerful techniques for inducing emotions (see Ekman 1983 et al.; Bloch et al. 1987) allow one to overcome the difficulty up to now encountered in inducing a positive mood. Furthermore, these techniques permit the induction of other emotions, in addition to sadness and happiness, such as tenderness, anger, fear, and sexual emotions. The joint use of emotion-inducing techniques and of the interference technique should greatly improve our understanding of the relative roles of the two hemispheres in the genesis of emotions.

Acknowledgements. The research described in this review was supported by the Italian CNR, Gruppo Nazionale di Neuroscienze and by MPI.

REFERENCES

Anzola GP, Bertoloni G, Buchtel HA, Rizzolatti G (1977) Spatial compatibility and anatomical factors in simple and choice reaction time. Neuropsychologia 15: 295-302

Bashore TR (1981) Vocal and manual reaction time estimates of interhemispheric transmission time. Psychol Bull 89: 352-368

Bear DM (1983) Hemispheric specialization and the neurology of emotion. Arch Neurol 40: 195-202

Berlucchi G, Crea F, Di Stefano M, Tassinari G (1977) Influence of spatial stimulus response compatibility on reaction time of ipsilateral and

contralateral hand to lateralized light stimuli. J Exp Psychol °Hum Perceptè 3: 505-517
Bloch S, Orthous P, Santibanez G (1987) Effector patterns of basic emotions: a psychophysiological method for training actors. J Soc Biol Struct 10: 1-19
Davidson RJ, Schwartz GE, Seron C, Bennett J, Galeman DJ (1979) Frontal versus parietal EEG asymmetry during positive and negative affects. Psychophysiology 16: 202
Donders FC (1869) Over de snelheid van psychische processen. Nederlandsch Archief voor Genees-en Natuurkunde 4: 117-145 Translation: Attention and performance II. Koster WG (ed) North-Holland, Amsterdam, 1969, pp 412-431
Ekman P, Levenson RW, Friesen WV (1983) Autonomic nervous system activity distinguishes among emotions. Science 221: 1208-1210
Fromm-Auch D (1982) Comparison of unilateral and bilateral ECT: evidence for selective memory impairment. Br J Psychiatry 141: 608-613
Gainotti G (1972) Emotional behavior and hemispheric side of the lesion. Cortex 8: 41-55
Hamilton M (1960) A rating scale for depression. J Neurol Neurosurg Psychiatry 23: 56-62
Jeeves MA, Dixon NF (1970) Hemispheric differences in response rates to visual stimuli. Psychon Sci 20: 249-251
Kahneman D (1973) Attention and effort. Prentice-Hall, Englewood Cliffs, New Jersey
Kronfol Z, Hamsher KD, Digre K, Waiziri R (1978) Depression and hemispheric function: changes associated with unilateral ECT. Br J Psychiatry 132: 560-567
Ladavas E, Nicoletti R, Rizzolatti G, Umilta' C (1984) Right hemisphere interference during negative affect: a reaction time study. Neuropsychologia 22: 479-485
Ley RG, Bryden MP (1979) Hemispheric differences in processing emotions and faces. Brain Lang 7: 127-138
Liotti M, Sava D, Caffarra P, Rizzolatti G (1988) Hemispheric differences in patients affected by depression and anxiety. Brain Cogn (in press)
Perris C, Monakov K (1979) Depressive symptomatology and systemic structural analysis of EEG. In: Cruzelier J, Flor-Henry P (eds) Hemisphere asymmetry of function in psychopathology. Elsevier, Amsterdam 81-98
Perris C, Monakov K, Von Knorring L, Botskarev V, Nikiforov A (1979) Systemic structural analysis of the electroencephalogram of depressed patients: general principles and preliminary results of an international collaborative study. Neuropsychology 4: 207-228
Poffenberger AT (1912) Reaction time to retinal stimulation with special reference to the time lost in conduction through nerve centers. Arch Psychol 23: 1-73
Rizzolatti G (1979) Interfield differences in reaction times to lateralized visual stimuli in normal subjects. In: Steele Russell I, Van Hof MW, Berlucchi G (eds) Structure and function of cerebral commissures. Macmillan, London, pp 390-399
Rizzolatti G, Bertoloni G, Buchtel MA (1979) Interference of concomitant

motor and verbal tasks on simple reaction time: a hemispheric difference. Neuropsychologia 17: 323-330

Rizzolatti G, Bertoloni G, De Bastiani PL (1982) Interference of concomitant tasks on simple reaction time: attentional and motor factors. Neuropsychologia 20: 447-455

Sava D, Liotti M, Rizzolatti G (1988) Right hemisphere superiority for programming oculomotion: evidence from simple reaction time experiments. Neuropsychologia (in press)

Sherrington CS (1906) The integrative action of the nervous system. Yale University Press, New Haven p 393

Tucker DM (1981) Lateral brain function, emotion and conceptualization. Psychol Bull 89: 19-46

Tucker DM, Stenzile CE, Roth RS, Shearer SL (1981) Right frontal lobe activation and right hemisphere performance. Arch Gen Psychiatry 38: 169-174

A Model of Adaptive Behavior Related to Cerebral Participation in Emotional Control

M. Kinsbourne

Department of Behavioral Neurology, Eunice Kennedy Shriver Center, Waltham, Massachusetts, USA, and Department of Neurology, Harvard Medical School, Boston, Massachusetts, USA

Adaptive Symptoms

The symptoms that arise from focal cerebral damage have been subdivided into the negative, the positive, and the compensatory (Jackson 1879). Negative symptoms represent the skill that is lost, and positive symptoms the effect of release of previously inhibited structures. Compensatory symptoms involve the attainment of the behavioral goal by other means. A fourth category should be added for consideration. This is adaptive symptomatology. It encompasses how the brain functions, given the cognitive loss, and the existential predicament that the loss poses for the individual. A dramatic instance derives from unilateral personal neglect. The loss of direct experience of the existence of one's left limbs conflicts irreconcilably with one's knowledge that they exist. Denying, negative and defensive attitudes about the left side of his body represent the patient's imperfect efforts to come to terms with (adapt to) an intractable contradiction. A similar consideration arises in emotional lateralization. It is overly simplistic to attribute negative emotion after left hemisphere damage to disinhibition of a "pessimistically disposed" right hemisphere or elation after right hemisphere impairment to disinhibition of an inherently "optimistic" left brain (Flor-Henry 1979). Novel concepts, like optimistic and pessimistic hemisphere, should only be introduced for phenomena that defy all attempts at explanation in terms of existing knowledge. It is more parsimonious, and perhaps more correct, to suppose that emotional alterations after focal cerebral damage express the reorganization of brain function in response to the selective depletion of the cognitive repertoire induced by the damage. Schilder (1950) remarked: "most of the symptoms we see in cortical lesions are not merely a defect but a changed activity, and this activity is the result of the parts of the brain which are still in function" (p. 328). When one takes into consideration that the brain has lost what is has lost it becomes understandable that the brain reacts in the way that it does.

In the account that follows, I do not treat emotions as constituting a separate domain from cognition, nor do I think the question, what comes first, the emotion (Zajonc 1980) or the cognition (Lazarus 1982) to be particularly pertinent. Instead, I regard emotions as interpolated at strategic points in the

Experimental Brain Research Series 18

basic blueprint of cognitive activity (the activity cycle). Emotions reflect the outcomes of actions, and actions are motivated by emotions.
Leont'ev (1978) has commented "emotions reflect relationships between motives (needs) and success, or the possibility of success, of realizing the action of the subject that responds to these motives" (p. 120). There will be emotional change if cognitive deficit impairs the prospects for success, and alternately, there will be suspension of cognitive effort if motivations cease to drive it. It follows that an emotional aberration referable to an area of damaged or dysfunctional brain could be:

1. A direct consequence of that dysfunction
2. A release phenomenon based on another previously inhibited area
3. Part of a compensatory effect
4. Adaptive to the potentialities and limitations of function of residual (undamaged) brain

Explanations by reorganization does not assume that the patient consciously appraises his loss of capability in activities of daily living, and generates an appropriate emotional reaction. Damage in each of the four quadrants of the cerebral cortex causes cognitive loss. But not all patients exhibit the expected distress, and when it occurs, the type of distress exhibited differs in ways not obviously related to the severity of the clinical incapacity. Adaptive reorganization is not merely a restatement of the obvious: that a patient cannot help but be distressed by the loss of a major capability. The emotional consequences of focal brain damaged defy pragmatic explanation. Having lost the ability to move a limb, one patient becomes depressed, another exhibits indifference. Losing language skill, one patient seeks to avoid the embarrassment of having to speak, another pours out unintelligible speech beyond any apparent necessity. The extent to which activities of daily living are impeded is an insufficient predictor of the emotional reaction. Nor has premorbid temperament been shown to explain the post- morbid emotional state. We have to take into account the topography of cerebral impairment to discover regularities in the pattern of patients' emotional response. At the highest level of mental function it is not loss of a facility that is in question, but rather the specifics of an impoverishment of some aspect of how the patient is able to relate to his world and to his own self.

Parallel to the literature on emotional concomitants of focal brain injury is the literature on neuropsychological change in affective disorder (see Gainotti 1979, in the context of his pioneering work in this field). An adequate theory of emotional representation at the cerebral level must encompass both data sets, preferably relying on the same underlying principles. In focal cerebral injury, a cognitive loss may be antecedent to an emotional response. But in psychopathology, the cognitive changes in question should in the first instance be assumed to be secondary to the disorder of affect. Only if this hypothesis has to be rejected can the stronger assumption be entertained: that the cognitive deficit identifies an area of primary cortical impairment that is also involved in the genesis of the psychopathology.

If emotional syndromes following lateralized brain change are simply expressions of the predominant functioning of the intact hemisphere, then many findings remain unexplained. In this discussion, I shall not again review the much reviewed findings with respect to emotional lateralization (e.g., Kinsbourne and Bemporad 1984; Silberman and Weingartner 1986). Instead, I shall complicate the issue by considering a subdivision of cortex into four quadrants (Finset 1983).This is, no doubt, still oversimplified, but it enables me to propose distinctions which appear to accommodate the bulk of findings to date. But my theory depends on certain assumptions about the function of these quadrants which, though not unreasonable, require further substantiation. The account that follows is offered for its explanatory value and does not purport to be known to be correct in detail.

Laterality and Caudality in Cerebral Specialization

Following Pribram and McGuinness (1975) I distinguish, in terms of "caudality," between posterior representational cortex and the anterior control systems (located in posterio- parietal and prefrontal-anterior-temporal cortex respectively). The representational area assembles a data base about the individual's internal state and his situation in ambient space. The control systems organize action on the internal state and the environment that has the goal of approximating these states to the adaptively determined ideal of the moment. Both systems are lateralized in a manner which I now address.

Any characterization of hemisphere specialization which aims to encompass cognitive and affective variables must transcend such strictly cognitive dichotomies as verbal versus nonverbal, successive versus simultaneous or analytic versus holistic, although it should accommodate the valid distinctions implied by each of the many currently popular formulations. Relying on the notorious conservatism in its basic design of the vertebrate central nervous system I venture a formulation in terms of the most basic decision available to even the simplest motile organism: to approach or to withdraw (Kinsbourne 1982).

Approach can be viewed at various levels of abstraction as sequential behavior organized so as to approximate the individual to a point; be the point a physical location in space or, at a more abstract level, any fully specified goal within the organism's repertoire (not necessarily at a particular physical location). Thus beyond movement toward a target, it encompasses fluent and automatized skills in language, perception and action, as well as approach concomitant with such affects as friendliness and interest. In contrast, withdrawal implies an emptiness rather that a point and is concomitant with negative emotions such as fear, disgust, contempt and sadness. This distinction can be further elaborated, taking into consideration the ability of animals with more than rudimentary nervous systems to match outcome to expectation and, upon mismatch, to arrest ongoing behavior, gather more information, in the light of which to reformulate plans. In this sequence of events, approach corresponds to continuing ongoing behavior, whereas withdrawal corresponds to arrest of ongoing behavior. In Pribram and McGuiness' (1975) terms, approach corresponds to tonic <u>activation</u>, which in mammals appears to be dopamine-mediated, and withdrawal to, phasic <u>arousal</u>, which is norepinephrinergic. The two categories of behavior are coordinated in what I have called the <u>activity cycle</u>.

The Activity Cycle

The organism is observed in ongoing planned activity. The activity is motivated toward a goal. For as long as achievement of the goal remains uncertain, anxiety maintains attention on a narrow set of task-relevant cues. The planned activity consists of a series of acts. After each act the physical situation is monitored to confirm that it has been transformed in the predicted manner to bring the individual closer to his goal. If the goal is achieved, behavior is suspended in the subjective context of a feeling of satisfaction. But if at any stage there is mismatch between schema and monitored consequence, action is arrested in the context of a surge of negative emotion, while the individual broadens the basis of data collection so as to be able to formulate an alternative plan in view of the unexpected event and changed situation. If such a plan is feasible, emotional arousal is allayed and ongoing behavior resumes. if not, emotional arousal continues until a new goal is substituted, that is more appropriate to the new situation.

I attribute the ongoing behavior to the activity of specialized areas of the left hemisphere. Concomitant with this behavior, but not necessarily similarly localized, we subjectively experience motivation and anxiety. The left frontotemporal cortex formulates plans and the left posterior cortex represents the data base pertinent to the outcomes of the plans (Kolb and Milner 1981). During this time, the left hemisphere inhibits emotional responsivity triggered by the right (Tucker and Newman 1981). But a major mismatch (that is, one to correct which no action subroutine has already been formulated and readied for use) results in arrest of left hemisphere activity. The right posterior area is specialized to be sensitive to cues that facilitate the emotional arousal response that interrupts the activity. If it is determined that the goal cannot be attained by alternative means at the individual's disposal, the context-sensitive right frontal area is responsible for accommodating the organism's aspirations to reality. Such accommodation dampens emotional arousal swings to the point that motivated behavior may resume.

The function of the left frontal lobe in planning is well known (Luria 1973). Left posterior lesions impair ability to identify and act upon, external stimuli (varying from visual object agnosia and alexia to apraxia and agraphia). The specialization of the right hemisphere for novelty detection has been repeatedly remarked upon, notably by Goldstein and Costa (1982). Arousal deficit after right posterior lesion (Heilman et al. 1978), and especially to stimuli with emotional connotation (Morrow et al. 1981; Zoccolotti et al. 1982) is particularly well illustrated by the current findings of Caltagirone and his colleagues (this volume). This group has also produced the first empirical evidence of the predicted deficit in avoidance behavior after right hemisphere lesions. They found a general decrease in gaze aversions and a loss of the relationship between emotionally negative input and averted gaze in right but not left hemisphere lesioned patients.

The Activity Cycle and Emotional Lateralization

While there is general agreement that the right hemisphere is specialized for "negative" emotions, there is dispute as to whether positive emotions are right or left lateralized (Fox and Davidson 1984). My suggested dichotomy between left lateralized approach and right lateralized withdrawal lent itself to the latter interpretation (Schwartz et al. 1975). However, in view of the present model, the dichotomy can be refined, and somewhat modified.

Interruption of planned activity by "mismatch" would understandably generate negative feelings, attending the right hemisphere's interrupt function. But surprise, not necessarily negative in valence, would also be a consequence of right lateralized processing. On the other hand, activity completed as planned would leave the person satisfied or gratified, leading to a state of calm, attributable to left hemisphere functioning. In contrast, incomplete business of the left hemisphere will be attended by anxiety, as has already been commented upon. Again, the fact that left hemisphere activity determines whether the individual feels calm or anxious, does not imply that these feelings are themselves left-lateralized. There is no evidence that such feelings (or any feelings) arise directly from neural discharges in the left hemisphere.

With the set of positive emotions, a distinction should be made between elation and euphoria as manifestation of phasic emotional arousal, and optimistic contentment, as a state. The former would reflect a right hemispheric interrupt function, the latter a left hemispheric state generating successful or succeeding action.

It follows that we should substitute dual polarities, one subserved by each hemisphere, for the single functional polarity between hemispheres that is usually advanced. The appropriate ones seem to be those advocated by Tucker et al. (1986), derived from Tellegen (1985), who, based on a factor analytic study, distinguished an affective dimension ranging from sluggish to elated, and another from calm to anxious. I regard the former as secondary to right hemisphere function, the latter as secondary to left. By this reasoning, disorders of mood (inner directed) unipolar and bipolar, would be related to the cognitive in activity of the right hemisphere and disorders of attitude (outed directed) to the cognitive activity of the left hemisphere. But first I consider what light these interpretations might cast on affective change after focal cerebral injury.

Diagonal Pattern of Post-Stroke Depression

The above account of the activity cycle and its cortical basis permits hypotheses explaining the "diagonal" pattern of post-stroke emotion (depressive versus indifference) noted by Finset (1983).

Suspended functioning of left prefrontal lobe (for whatever reason) leaves the individual unable to plan for the future, or indeed to do more than react to changes in the immediate stimulus situation. The loss of ability to implement his goals would understandably make the individual feel helpless and vulnerable and, as it continues, hopeless. A wide range of environmental contingencies that could normally be easily handled are now out of control and therefore menacing to a varied extent. Confronted by challenge, the "catastrophic reaction" can result, which Gainotti (1972) explained as a normal (i.e., appropriate) response to the cognitive deficit. The patient who has lost access to the frontal planning facilities is not incapable of feeling pleasure when the environment accommodates his wishes, but, given the vicissitudes of everyday existence, his sense of ineffectiveness will soon return. Also, because motivated behaviors continue and yet goals are not met, the patient's depression will be attended by anxiety.

Depression consequent upon right posterior hypofunction should have different characteristics. The cognitive elaborations (expressions of helplessness and hopelessness) of left frontal depression would not be apparent. Instead, the patient feels emotionally numbed, neither enjoying what used to please, nor reacting much to negative events either. This pervasive anhedonia leaves the individual motiveless (because no goal seems worth striving for) and therefore also free of anxiety. Rather than cognitively elaborated the depression will more often express itself in somatic symptoms.

Severe depression might well involve underactivation of both left anterior and right posterior quadrants (especially when it is based on the chaotic pattern of cerebral metabolic function found in bipolar disease - Wood and Flowers 1988). As indicated by the activity cycle concept, suspended right posterior function would undermine the motivation for left frontal activity. Conversely, left frontal incapacity would be apt to set up recurring failures that could cause right hemisphere functional rearrangements as described. But some promise for a neuropsychological taxonomy of depression is offered by the symptomatic distinction between endogenous and reactive (or endogenomorphic and neurotic - Klein 1974) variants of depression. Anhedonia is basic to the former, and dysphoric irritability to the latter. It seems warranted to look for differential neuropsychological profiles in these two conditions. Cerebral metabolic findings could also be analyzed in this way (note the findings by Baxter et al. (1985) of left frontal hypofunction in a subset of depressives). Finally, differential treatment effects might be expected. Anhedonia should yield to norepinephrine agonists (and norepinephrine projections are allegedly more profuse to right hemisphere - Oke et al. 1978). Dysphoric irritability should respond to serotonin agonists (Liebowitz et al. 1984).

A quadrant perspective on brain state suggests a reconsideration of the voluminous literature on laterality in psychiatric disorders (Flor-Henry 1979). Different laterality tests are more likely to tap lateral imbalances in activation of posterior and of anterior cerebral cortex.

Laterality Indices of Cerebral Activation

Laterality paradigms can be subgrouped according to whether they tap input or output processes. Left-right balance dichotic listening and hemifield viewing reflect the state of posteriorly localized analyses for auditory and visual information respectively. In contrast, directional movement tendencies reflect anterior patterns of cortical activation. These include directional bias in reflective gaze (Kinsbourne 1972) locomotor circling, differential hand use in gesturing (Kimura 1973) and perhaps even hand preference.

According to Kimura's (1961) original "structural" model for laterality effects, these directly index hard-wired hemisphere specialization. If so, no changes in such effects should accompany psychopathology, as the idea that the cortex changes its hard-wiring when psychiatric disease strikes is too far fetched to be seriously considered. Even more remote is the possibility that the hard-wiring varies in relation to affective state, laterality being considered a state marker (Kronfol et al. 1978; Brumback et al. 1980), let alone fluctuate on a daily basis (Wexler 1988). To make sense of laterality in psychopathology one has to enlist the concept of selective hemisphere activation (Kinsbourne 1970). This intervening variable accounts for the task-relatedness of laterality effects; tasks that enlist left hemisphere resources bias attention rightwards, and vice versa. But it is also open to other influences which can modify, override and even reverse the expected task-related perceptual asymmetry. Tonic changes in the activation of local cortical areas may interact with task-related activation to lead to a vector resultant of laterally directed attention that diverges from expectation. Such tonic changes may reflect mood changes (or even individual variation in the general population - Levy et al. 1983). For purposes of explaining laterality in affective disorder, I consider hemispheric neural tissue as costituting a multipurpose computational space (Kinsbourne 1987) in which the same or interacting areas are involved both in emotionally relevant processing and the control of lateral attention. If such an area becomes inactive (or overactive) in relation to this role in the control of emotions, it will necessarily also become under- or over influential in the control of lateral attention. According to this concept, biases in the direction of lateral attention (i.e., laterality effects) are markers for the balance of activation between the cerebral hemispheres, and thus potentially reflect lateralized activation changes.

Reports of opposing perceptual asymmetries in bipolar patients depending on whether they are in the manic or the depressive phase (Sackeim et al. 1983) and even of opposing hand preference (Bruce 1895; Flor-Henry 1979) would be further cases in point.

If laterality effects reflect the balance of activation between the hemispheres, a further prediction follows. Laterality effects for input decoding tasks that, according to the functional cerebral distance principle (Kinsbourne and Hicks 1978), most directly tap the activation state of the posterior cerebral cortex, should be most sensitive to changes in activation of one or other posterior quadrant. Output laterality tests should more closely reflect the activational balance of the frontal lobes.

The phenomenon of locomotor "circling," commonly observed after unilateral ablation at several levels of the mammalian brain, provides an illustration of how the balance of activation between laterally symmetrical brain facilities underlies laterality phenomena. Circling clockwise cannot simply be understood as a bias in motor directionality. Consider the following observation. Within the confines of a box, a unilaterally lesioned animal circles clockwise. Outside its periphery, it circles counterclockwise (Wise and Holmes 1986).
These behavior patterns can be understood as due to a lesion-induced bias when the animal orients to the wall of the box. The animal does not veer off into space. It confronts the lateraly extended stimulus, but on account of the lesion, orients not centrally but toward its right side. The disinhibited rightward orienting tendency of the intact left brain controls behavior. Right sided stimuli become "magnets" for the animal's attention. As the wall of the box demarcates an enclosed space, there is no boundary at which this biased orienting can stop. It therefore continues, in principle indefinitely.

There is voluminous evidence that lateral orienting is controlled by an opponent system across cerebral hemispheres (reviewed by Kinsbourne 1974) and that activating one side (by electrical stimulation or an anticholinesterase), just as inactivating (excising) the other, generates the bias controlateral to the more activated side.

Now consider the recent observation that on free field, schizophrenics circle counterclockwise (Bracha 1987). This suggests an imbalance at some relevant level of the brain, probably prefrontal cortex (Crowne and Pathna 1982), such that the left is less in control of behavior than the right. The prolonged latency to identification of briefly exposed stimuli in the right visual field in schizophrenics (Connolly et al. 1979), and Posner et al. (1988) observation of impaired rightward attentional shifts to unexpected locations by schizophrenics represent the same basic phenomenon.

According to the activity cycle model, depressed patients (especially reactive depressives) should also circle counterclockwise and patients with indifference syndrome clockwise.

Laterality findings in depression are consistent with the above neuropsychological subtyping of depression. Bruder, Quitkin, Stewart et al. (in press) found enhanced right ear advantage in dichotic listening for melancholics, for whom anhedonia was defining characteristic. This would be predicted if there is right posterior hypofunction. But a group of patients with atypical depression, who retain reactivity of mood but exhibit extreme inertia, yielded normal dichotic laterality. A left frontal hypofunction would not be expected to affect dichotic laterality.

An as yet untested prediction arises. Output laterality should yield the opposite pattern - leftward biased asymmetries in the atypical group, but normal findings in the melancholics.

Indifference syndromes should also subdivide into contrasting patterns, depending on whether the hypofunction is primarily left posterior or right anterior. With left posterior underactivation, the subject's ability to monitor the impact of his actions on the environment is limited (at the level of interpretation). Specifically, he will fail to pick up mismatch signals to the effect that what he was doing was not having the desired effect. A striking example derives from easily observed behavior of the patient with a posterior (Wernicke) aphasia. Although his speech is largely unintelligible, he does not hesitate to use it freely to express his needs. When the listener shows by means of facial expressions and other actions that he has not understood, the patient fails to realize that he is not effectively communicating. Instead he may become angry on account of the listener's imagined noncompliance. This type of emotional reaction is better described as oblivious than indifferent.

In contrast, the patient with right frontal hypofunction has no illusions about the consequences of his actions on others, but does not let these interfere with what he wants to do (Luria 1973). He no longer compares his feelings to those of other people and his behavior becomes impulsive and disregards the interest of others and the conventions of groups. Information which would normally elicit a strong emotional reaction, be it about relatives or about him himself, is brushed off as unimportant. The term indifference is apt for the syndrome of right frontal underactivity.

Evidence from Starkstein and Robinson (1988) and Finset (1988) relating the quadrant affected by stroke to the presence of a depressive versus an indifference syndrome might lend itself to the above interpretations. If depressive syndromes appear to be more severe after left frontal than right posterior stroke in these investigators' data, this may simply be because the Hamilton Depression questionnaire they used happens to incorporate more entries sensitive to the left frontal type. The "indifference" reactions attributed to left posterior and right anterior stroke have not been rated and item-analyzed, but with respect to depression, Finset (1988) has reviewed his data and finds support for the above formulation.

Neuropsychological Tests in Depression

Neuropsychological testing of depressed patients most often indicates selective difficulty with tasks that are thought to reflect the functioning of the right posterior parietal cortex. In contrast, where cerebral metabolic measure have been found to be abnormal in unipolar depression, this abnormality is typically left frontal in distribution (Baxter et al. 1985). It may be that the two proposed subtypes distributed differently across different study samples.

There is some encouragement in the psychiatric literature for the notion of two variants of depressive symptomatology, considered as at opposite poles of a continuum (Paykel et al. 1974). They may correspond to opposing balances between norepinephrine and serotonin activity (Gray 1982).
Norepinephrine activity may be high and serotonin low in reactive depression,

and the reverse in endogenous depression (Otto et al. 1987). Patients with endogenous depression are characterized by unrelenting anhedonia and somatic complaints. In their families, other cases of unipolar depression are to be found. In reactive (or neurotic) depression, frequent bouts of negative affect result in relation to adversities which these people would seem to bring upon themselves. They are able to respond to pleasurable circumstances but anxiety levels are high. In their family histories, attention deficit disorder, alcholism, and bipolar depression are commonly found. In our own ongoing study of childhood and adolescent depressives, analogous patterns seem to emerge: A subgroup of "pure depressives" with depression in their family histories and particular difficulty on tests of right posterior localizing value and a subgroup of combined depression and attention deficit, with a more varied set of characterologic problems in the family and most striking difficulties in frontal tests.

If depressive subtypes do indeed relate to hypofunction of obliquely opposite cerebral quadrants, this does not therefore prove a casual relationship between psychiatric symptomatology and the cortical localizations. For istance, findings that take this form should not be construed as showing that affective disorder is caused by malfunction of cerebral areas. Whereas an individual's cerebral equipment is genetically determined, the use to which he puts it is presumably affected by environmental variables. The person might well suspend use of parts of his brain for reasons that lend themselves better to explanation in terms of contingencies or dynamic factors than ones which are neuropsychological. We should view the role of neuropsychology in psychopathology as contributing toward a map of how the patient uses his cerebral cortex, depending on the nature of the mental disorder and its clinical state.

Five decades ago, Paul Schilder (1935) remarked: "The old neuropathology believed that the functions are localized in the parts of the brain, the lesion of which disturbs the function. We know today that this is not so. Particular parts of the brain are necessary for the function, but the psychic function does not have its seat in these parts (p. 327). Curiously, it appears we no longer know what we knew then. Theorists once again "localize" functions in brain parts, lesions of which disturb those functions. This ill considered practice appears at its most absurd when investigators suppose that, if one is sad after a particular locus of lesion, that locus was specialized to stop us from being sad, or to preclude another part from generating sadness (and vice versa for happiness). In this discussion, I have attempted to show that one can explain the appearance of particular emotional reactions after specific focal cerebral lesion without have to "localize" the corresponding emotions in those areas of cerebral cortex. Instead, the cognitive rearrangement, necessitated by the lesion-induced, deficits, elicit the emotional changes in question. The latter can be attributed to limbic activity, as has been long supposed, and for good reason (p 1955; Papez 1937).

REFERENCES

Baxter LR, Phelps ME, Mazziotta JC, et al (1985) Cerebral metabolic rates for glucose in mood disorder: studies with positron emission tomography and flurodeoxyglucose F18. Arch Gen Psychiatry 42: 441-447

Bracha HS (1987) Asymmetric rotational (circling) behavior, a dopamine-related asymmetry: preliminary findings in unmedicated and never-medicated schizophrenic patients. Biol Psychiatry 22: 995-1003

Bruce LC (1895) Notes of a case of dual brain action. Brain 18: 54-65

Bruder G, Quitkin F, Stewart JW, McGrath P, Martin C, Voglomaier M, Harrison W (submitted) Cerebral laterality and depression: Perceptual asymmetry in diagnostic subtypes before and after antidepressant treatment.

Brumback RA, Staton RD, Wilson H (1980) Neuropsychological study of children during and after remission of endogenous depressive episodes. Percept Mot Skills 50: 1163-1167

Connelly JF, Gruzelier HH, Kleinman DM, et al (1979) Lateralized abnormalities in hemisphere-specific tasks in psychiatric patients and controls. In: Gruzelier J, Flor-Henry P (eds) Hemisphere asymmetries of function in psychopathology. Elsevier-North Holland Biomedical, New York

Crowne DP, Pathnia MN (1982) Some attentional effects of unilateral frontal lesions in the rat. Behav Brain Res 6: 25-39

Finset A (1983) Depressive behavior, outburst crying, and emotional indifference in left hemiplegics. Paper presented at the second annual symposium of models and techniques of cognitive rehabilitation, Indianapolis

Finset A (1988) Depressed mood and reduced emotionality after right-hemisphere brain damage. In: Kinsbourne M (ed) Cerebral hemisphere function in depression. American Psychiatric Press, Washington, D.C.

Flor-Henry P (1979) On certain aspects of the localization of the cerebral systems regulating and determining emotion. Biol Psychiatry 14: 677-698

Fox NA, Davidson RJ (1984) Hemisphere substrates of affect: a developmental model. In: Fox NA, Davidson RJ (eds) Psychobiology of affective development. Erlbaum, Hillsdale

Gainotti G (1972) Emotional behavior and hemispheric side of the lesion. Cortex 8: 41-55

Gainotti G (1979) The relationships between emotions and cerebral dominance. In: Gruzelier J, Flor-Henry P. (eds) Hemisphere asymmetries of function in psychopathology. Elsevier, New York

Gray JA (1982) Precis of the neuropsychology of anxiety: an inquiry into the functions of the septo-hippocampal system. Behav Brain Sci 5: 469-534

Heilman KM, Schwartz HD, Watson RT (1978) Hypoarousal in patients with the neglect syndrome and emotional indifference. Neurology 28: 229-232

Jackson JH (1879) On affectations of speech from disease of the brain. Brain 2: 203-222

Kimura D (1961) Cerebral dominance and the perception of verbal stimuli. Can J Psychol 15: 166-171

Kimura D (1973) Manual activity during speaking-I. right-handers. Neuropsychologia 11: 45-50

Kinsbourne M (1970) The cerebral basis of lateral asymmetries in attention. Acta Psychol 33: 193-201
Kinsbourne M (1972) Eye and head-turning indicates cerebral lateralization. Science 176: 539-541
Kinsbourne M (1974) Lateral interactions in the brain. In: Kinsbourne M, Smith WL (eds) Hemispheric disconnection and cerebral function. Thomas, Springfield
Kinsbourne M (1982) Hemispheric specialization and the growth of human understanding. Am Psychol 37: 411-420
Kinsbourne M (1987) The material basis of mind. In: Vaina L (ed) Matters of intelligence. Reidel, Boston
Kinsbourne M, Bemporad B (1984) Lateralization of emotion/ a model and the evidence. In: Fox N, Davidson RJ (eds) The psychobiology of affective development. Erlbaum, Hillsdale
Kinsbourne M, Hicks RE (1978) Functional cerebral space: a model for the overflow, transfer and interference effects in human performance: A tutorial review. In: Requin J (ed) Attention and performance VII. Erlbaum, Hillsdale
Klein DF (1974) Endogenomorphic depression. Arch Gen Psychiatry 31: 447-455
Kolb B, Milner B (1981) Performance of complex arm and facial movements after focal brain lesions. Neuropsychologia 19: 491-503
Kronfol Z, Hamsher K, Digre K, Wazir R (1978) Depression and hemisphere functions: Changes associated with unilateral ECT. Br J Psychiatry 132: 560-567
Lazarus R (1982) Thoughts on the relations between emotion and cognition. Am Psychol 37: 1019-1024
Leont'ev AN (1978) Activity, consciousness and personality. Prentice-Hill, Englewood Cliffs
Levy J, Haller W, Banich MT et al (1983) Are variations among right-handed individuals in perceptual asymmetries caused by characteristic arousal differences between hemispheres? J Exp Psychol Human Percept 9: 329-358
Liebowitz MR, Quitkin FM, Stewart JW, McGrath PJ, Harrison W, Rabkin J, Tricano E, Markowitz JS, Klein DF (1984) Phenelzine v Imipramine in atypical depression: a preliminary report. Arch Gen Psychiatry 120: 669-677
Luria AR (1973) The working brain: an introduction to neuropsychology. New York, Basic
MacLean PD (1955) The limbic system ("visceral brain") in relation to central gray and reticulum of the brain stem. Psychosom Med 17: 355-366
Morrow L, Vrtunski PB, Kim Y, Boller F (1981) Arousal responses to emotional stimuli and laterality of lesion. Neuropsychologia 19: 65-71
Oke A, Keller R, Meffort I et al (1978) Lateralization of norepinephrine in human thalamus. Science 200: 1141-1143
Otto MW, Yeo RA, Dougher MJ (1987) Right hemisphere involvement in depression: toward a neuropsychological theory of negative affective experiences. Biol Psychiatry 22: 1201-1215
Papez JW (1937) A proposed mechanism of emotion. Arch Neurol Psychiatry 38: 725-743
Paykel ES, Prusoff B, Klerman GL (1974) The endogenous-neurotic continuum in depression. J Psychiatr Res 8: 73-90

Posner MI, Easly TS, Reiman E, Pardo PJ, Dhawan M (1988) Asymmetry in hemispheric control of attention in schizophrenia. Arch Gen Psychiatry, 45: 814-821

Pribram KH, McGuinness D (1975) Arousal, activation, and effort in the control of attention. Psychol Rev 82: 116-149

Sackheim HA, Decina P, Epstein D, Bruder GE, Malitz S (1983) Possible reversed affective lateralization in a case of bipolar disorder. Am J Psychiatry 140: 9

Schilder P (1935) The image and appearance of the human body. International Universities Press, New York

Schwartz GE, Davidson RJ, Maier F (1975) Right hemisphere lateralization for emotion in the human brain: Interactions with cognition. Science 190: 286-288

Silberman EK, Weingartner H (1986) Hemispheric lateralization of functions related to emotion. Brain Cognition 5: 322-353

Starkstein SE, Robinson RG (1988) Lateralized emotional response following stroke. In: Kinsbourne M (ed) Cerebral hemisphere function in depression. American Psychiatric Press, Washington, D.C.

Tellegen A (1985) Structures of mood and personality and their relevance to assessing anxiety, with an emphasis on self-report In: Tuma AH, Maser JD (eds) Anxiety and the anxiety disorders. Erlbaum, Hillsdale

Tucker DM, Newman JP (1981) Verbal versus imaginal cognitive strategies in the inhibition of emotional arousal. Cognit Ther Res 5: 197-202

Wexler BE (1988) Regional brain dysfunction in depression. In: Kinsbourne M (ed) Cerebral hemisphere function in depression. American Psychiatric Press, Washington, D.C.

Wise, Holmes (1986) Brain Research Bulletin 16: 267

Wood FB, Flowers L (1988) Cortical activation in psychiatric disorder: A review of work in progress. In: Kinsbourne M (ed) Cerebral hemisphere function in depression. American Psychiatric Press, Washington, D.C.

Zajonc RB (1980) Feeling and thinking: preferences need no inferences. Am Psychol 35: 151-175

Zoccolotti P, Scabini D, Violani C (1982) Electrodermal responses in patients with unilateral brain damage. J Clin Neuropsychol 4: 143-150

Subject Index

MIX
Papier aus verantwortungsvollen Quellen
Paper from responsible sources
FSC® C105338

If you have any concerns about our products,
you can contact us on
ProductSafety@springernature.com

In case Publisher is established outside the EU,
the EU authorized representative is:
Springer Nature Customer Service Center GmbH
Europaplatz 3, 69115 Heidelberg, Germany

Printed by Libri Plureos GmbH
in Hamburg, Germany